D0543060

SPECIAL NEEDS IN ORDINARY SC

General Editor Peter Mittler

Science for All

Special Needs in Ordinary Schools

General editor: Peter Mittler
Associate editors: James Hogg, Peter Pumfrey, Tessa Roberts, Colin Robson
Honorary advisory board: Neville Bennett, Marion Blythman, George Cooke, John Fish, Ken Jones, Sylvia Phillips, Klaus Wedell, Phillip Williams

Science for All:
Teaching Science in the Secondary School

David J. Reid and Derek Hodson

Cassell

Cassell Educational Limited
Artillery House
Artillery Row
London SW1P 1RT

British Library Cataloguing in Publication Data

Reid, David J.
  Science for all: teaching science in the
  secondary school—(Special needs in ordinary schools)
  1. Slow learning children—Education—Science
  I. Title   II. Hodson, Derek   III. Series
  371.92′6   LC4707.5

ISBN: 0 – 304 – 31382 –3

Typeset by Activity Ltd., Salisbury, Wilts.
Printed and bound in Great Britain by Biddles Ltd.,
Guildford and King's Lynn

Last digit is print no: 9 8 7 6 5 4 3 2 1

Contents

Foreword: Towards education for all

AIMS

This series aims to support teachers as they respond to the challenge they face in meeting the needs of all children in their school, particularly those identified as having special educational needs.

Although there have been many useful publications in the field of special educational needs during the last decade, the distinguishing feature of the present series of volumes lies in their concern with specific areas of the curriculum in primary and secondary schools. We have tried to produce a series of conceptually coherent and professionally relevant books, each of which is concerned with ways in which children with varying levels of ability and motivation can be taught together. The books draw on the experience of practising teachers, teacher trainers and researchers and seek to provide practical guidelines on ways in which specific areas of the curriculum can be made more accessible to all children. The volumes provide many examples of curriculum adaptation, classroom activities, teacher–child interactions, as well as the mobilisation of resources inside and outside the school.

The series is organised largely in terms of age and subject groupings, but three 'overview' volumes have been prepared in order to provide an account of some major current issues and developments. Seamus Hegarty's *Meeting Special Needs in Ordinary Schools* gives an introduction to the field of special needs as a whole, whilst Sheila Wolfendale's *Primary Schools and Special Needs* and John Sayer's *Secondary Schools for All?* address issues more specifically concerned with primary and secondary schools respectively. We hope that curriculum specialists will find essential background and contextual material in these overview volumes.

In addition, a section of this series will be concerned with examples of obstacles to learning. All of these specific special needs can be seen on a continuum ranging from mild to severe, or from temporary and transient to long-standing or permanent. These include difficulties in learning or in adjustment and behaviour, as well as problems resulting largely from sensory or physical impairments or from difficulties of communication from whatever cause. We hope that teachers will consult the volumes in this section for guidance on working with children with specific difficulties.

The series aims to make a modest 'distance learning' contribution to meeting the needs of teachers working with the whole range of pupils with special educational needs by offering a set of resource materials relating to specific areas of the primary and secondary curriculum and by suggesting ways in which learning obstacles, whatever their origin, can be identified and addressed.

We hope that these materials will not only be used for private study but be subjected to critical scrutiny by school-based inservice groups sharing common curricular interests and by staff of institutions of higher education concerned with both special needs teaching and specific curriculum areas. The series has been planned to provide a resource for LEA advisers, specialist teachers from all sectors of the education service, educational psychologists, and teacher working parties. We hope that the books will provide a stimulus for dialogue and serve as catalysts for improved practice.

It is our hope that parents will also be encouraged to read about new ideas in teaching children with special needs so that they can be in a better position to work in partnership with teachers on the basis of an informed and critical understanding of current difficulties and developments. The goal of 'Education for All' can only be reached if we succeed in developing a working partnership between teachers, pupils, parents, and the community at large.

ELEMENTS OF A WHOLE-SCHOOL APPROACH

Meeting special educational needs in ordinary schools is much more than a process of opening school doors to admit children previously placed in special schools. It involves a radical re-examination of what all schools have to offer all children. Our efforts will be judged in the long term by our success with children who are already in ordinary schools but whose needs are not being met, for whatever reason.

The additional challenge of achieving full educational as well as social integration for children now in special schools needs to be seen in the wider context of a major reappraisal of what ordinary schools have to offer the pupils already in them. The debate about integration of handicapped and disabled children in ordinary schools should not be allowed to overshadow the movement for curriculum reform in the schools themselves. If successful, this could promote the fuller integration of the children already in the schools.

If this is the aim of current policy, as it is of this series of unit texts, we have to begin by examining ways in which schools and school policies can themselves be a major element in children's difficulties.

Can schools cause special needs?

Traditionally, we have looked for causes of learning difficulty in the child. Children have been subjected to tests and investigations by doctors, psychologists and teachers with the aim of pin-pointing the nature of their problem and in the hope that this might lead to specific programmes of teaching and intervention. We less frequently ask ourselves whether what and how we teach and the way in which we organise and manage our schools could themselves be a major cause of children's difficulties. Questions concerned with access to the curriculum lie at the heart of any whole-school policy. What factors limit the access of certain children to the curriculum? What modifications are necessary to ensure fuller curriculum access? Are there areas of the curriculum from which some children are excluded? Is this because they are thought 'unlikely to be able to benefit'? And even if they are physically present, do they find particular lessons or activities inaccessible because textbooks or worksheets demand a level of literacy and comprehension which effectively prevents access? Are there tasks in which children partly or wholly fail to understand the teacher's language? Are some teaching styles inappropriate for individual children?

Is it possible that some learning difficulties arise from the ways in which schools are organised and managed? For example, what messages are we conveying when we separate some children from others? How does the language we use to describe certain children reflect our own values and assumptions? How do schools transmit value judgements about children who succeed and those who do not? In the days when there was talk of comprehensive schools being 'grammar schools for all', what hope was there for children who were experiencing significant learning difficulties? And even today, what messages are we transmitting to children and their peers when we exclude them from participation in some school activities? How many children with special needs will be entered for the new General Certificate of Secondary Education examinations? How many have taken or will take part in Technical and Vocational Education Initiative schemes?

The argument here is not that all children should have access to all aspects of the curriculum. Rather it is a plea for the individualisation of learning opportunities for all children. This requires a broad curriculum with a rich choice of learning opportunities designed to suit the very wide range of individual needs.

Curriculum reform

The last decade has seen an increasingly interventionist approach by Her Majesty's Inspectors of Education, by officials of the Department of Education and Science and by individual Secretaries of State. The Great Debate, allegedly beginning in 1976, led to a flood of curriculum guidelines from the centre. The garden is secret no longer. Whilst Britain is far from the centrally imposed curriculum found in some other countries, government is increasingly insisting that schools must reflect certain key areas of experience for all pupils, and in particular those concerned with the world of work (sic), with science and technology, and with economic awareness. These priorities are also reflected in the prescriptions for teacher education laid down with an increasing degree of firmness from the centre.

There are indications that a major reappraisal of curriculum content and access is already under way and seems to be well supported by teachers. Perhaps the best known and most recent examples can be found in the series of ILEA reports concerned with secondary, primary and special education, known as the Hargreaves, Thomas and Fish Reports (ILEA, 1984, 1985a, 1985b). In particular, the Hargreaves Report envisaged a radical reform of the secondary curriculum, based to some extent on his book *Challenge for the Comprehensive School* (Hargreaves, 1982). This envisages a major shift of emphasis from the 'cognitive–academic' curriculum of many secondary schools towards one emphasising more personal involvement by pupils in selecting their own patterns of study from a wider range of choice. If the proposals in these reports were to be even partially implemented, pupils with special needs would stand to benefit from such a wholesale review of the curriculum of the school as a whole.

Pupils with special needs also stand to benefit from other developments in mainstream education. These include new approaches to records of achievement, particularly 'profiling', and a greater emphasis on criterion-referenced assessment. What about the new training initiatives for school leavers and the 14–19 age group in general? Certainly, the pronouncements of the Manpower Services Commission emphasise a policy of provision for all, and have made specific arrangements for young people with special needs, including those with disabilities. In the last analysis, society and its institutions will be judged by their success in preparing the majority of young people to make an effective and valued contribution to the community as a whole.

A CLIMATE OF CHANGE

Despite the very real and sometimes overwhelming difficulties faced by schools and teachers as a result of underfunding and professional unrest, there are encouraging signs of change and reform which, if successful, could have a significant impact not only on children with special needs but on all children. Some of these are briefly mentioned below.

First, we are more aware of the need to confront issues concerned with civil rights and equal opportunities. All professionals concerned with human services are being asked to examine their own attitudes and practices and to question the extent to which these might unwittingly or even deliberately discriminate unfairly against some sections of the population.

We are more conscious than ever of the need to take positive steps to promote the access of girls and women to full educational opportunities. We have a similar concern for members of ethnic and religious groups who have been, and still are, victims of discrimination and restricted opportunities for participation in society and its institutions. It is no accident that the title of the Swann Report on children from ethnic minorities was *Education for All*. This, too, is the theme of the present series and the underlying aim of the movement to meet the whole range of special needs in ordinary schools.

Special needs and social disadvantages

Problems of poverty and disadvantage are common in families of children with special needs already in ordinary schools. The probability of socially disadvantaged children being identified as having special needs is very much greater than for other children. Children with special needs are therefore doubly vulnerable to underestimation of their abilities – first, because of their family and social backgrounds and, second, because of their low achievements. A recent large-scale study of special needs provision in junior schools suggests that, although teachers' attitudes to low-achieving children are broadly positive, they are pessimistic about the ability of such children to derive much benefit from increased special needs provision (Croll and Moses, 1985).

Partnership with parents

The Croll and Moses survey of junior school practice confirms that teachers still tend to attribute many children's difficulties to adverse home circumstances. How many times have we heard comments along the lines of 'What can you expect from a child from that kind of family?' Is this not a form of stereotyping at least as damaging as racist and sexual attitudes?

Partnership with parents of socially disadvantaged children thus presents a very different challenge from that portrayed in the many reports of successful practice in some special schools. Nevertheless, the challenge can be and is being met. Paul Widlake's recent books (1984, 1985) give the lie to the oft-expressed view that some parents are 'not interested in their child's education'. Widlake documents project after project in which teachers and parents have worked well together. Many of these projects have involved teachers visiting homes rather than parents attending school meetings. There is also now ample research to show that children whose parents listen to them reading at home tend to read better and to enjoy reading more than other children (Topping and Wolfendale, 1985; see also Sheila Wolfendale's *Primary Schools and Special Needs*, in the present series).

Support in the classroom

If teachers in ordinary schools are to identify and meet the whole range of special needs, including those of children currently in special schools, they are entitled to support. Above all, this must come from the head teacher and from the senior staff of the school; from any special needs specialists or teams already in the school; from members of the new advisory and support services, as well as from educational psychologists, social workers and any health professionals who may be involved.

This support can take many forms. In the past, support meant removing the child for considerable periods of time into the care of remedial teachers either on the school staff or coming in from outside. Withdrawal now tends to be discouraged, partly because it is thought to be another form of segregation within the ordinary school, and therefore in danger of isolating and stigmatising children, and partly because it deprives children of access to lessons and activities available to other children.

We can think of the presence of the specialist teacher as being on a continuum of visibility. A 'high-profile' specialist may sit alongside a pupil with special needs, providing direct assistance and support in participating in activities being followed by the rest of the class. A 'low-profile' specialist may join with a colleague in what is in effect a team-teaching situation, perhaps spending a little more time with individuals or groups with special needs. An even lower profile is provided by teachers who may not set foot in the classroom at all but who may spend considerable periods of time in discussion with colleagues on ways in which the curriculum can be made more accessible to all the children in the class, including the least able. Such discussions may involve an examination of textbooks and other reading assignments for readability, conceptual difficulty and relevance of content, as well as issues concerned with the presentation of the material, language modes and complexity used to explain what is required, and the use of different approaches to teacher–pupil dialogue.

IMPLICATIONS FOR TEACHER TRAINING

Issues of training are raised by the authors of the three overview works in this series but permeate all the volumes concerned with specific areas of the curriculum or specific areas of special needs.

The scale and complexity of changes taking place in the field of special needs and the necessary transformation of the teacher-training curriculum imply an agenda for teacher training that is nothing less than retraining and supporting every teacher in the country in working with pupils with special needs.

Whether or not the readers of these books are or will be experiencing a training course, or whether their training consists only of the reading of one or more of the books in this series, it may be useful to conclude by highlighting a number of challenges facing teachers and teacher trainers in the coming decades.

1. We are all out of date in relation to the challenges that we face in our work.
2. Training in isolation achieves very little. Training must be seen as part of a wider programme of change and development of the institution as a whole.
3. Each LEA, each school and each agency needs to develop a strategic approach to staff development, involving detailed identification of training and development needs with the staff as a whole and with each individual member of staff.
4. There must be a commitment by management to enable the staff member to try to implement ideas and methods learned on the course.
5. This implies a corresponding commitment by the training institutions to prepare the student to become an agent of change.
6. There is more to training than attending courses. Much can be learned simply by visiting other schools, seeing teachers and other professionals at work in different settings and exchanging ideas and experiences. Many valuable training experiences can be arranged within a single school or agency,

or by a group of teachers from different schools meeting regularly to carry out an agreed task.
7. There is now no shortage of books, periodicals, videos and audio-visual aids concerned with the field of special needs. Every school should therefore have a small staff library which can be used as a resource by staff and parents. We hope that the present series of unit texts will make a useful contribution to such a library.

The publishers and I would like to thank the many people – too numerous to mention – who have helped to create this series. In particular we would like to thank the Associate Editors, James Hogg, Peter Pumfrey, Tessa Roberts and Colin Robson, for their active advice and guidance; the Honorary Advisory Board, Neville Bennett, Marion Blythman, George Cooke, John Fish, Ken Jones, Sylvia Phillips, Klaus Wedell and Phillip Williams, for their comments and suggestions; and the teachers, teacher trainers and special needs advisers who took part in our information surveys.

Professor Peter Mittler   University of Manchester
                                                        January 1987

REFERENCES

Croll, P. and Moses, D. (1985) *One in Five: The Assessment and Incidence of Special Educational Needs*. London: Routledge & Kegan Paul.

Hargreaves, D. (1982) *Challenge for the Comprehensive School*. London: Routledge & Kegan Paul.

Inner London Education Authority (1984) *Improving Secondary Education*. London: ILEA (The Hargreaves Report).

Inner London Education Authority (1985a) *Improving Primary Schools*. London: ILEA (The Thomas Report).

Inner London Education Authority (1985b) *Educational Opportunities for All?* London: ILEA (The Fish Report).

Topping, K. and Wolfendale, S. (eds.) (1985) *Parental Involvement in Children's Reading*. Beckenham: Croom Helm.

Widlake, P. (1984) *How to Reach the Hard to Teach*. Milton Keynes: Open University Press.

Widlake, P. (1985) *Reducing Educational Disadvantage*. London: Routledge & Kegan Paul.

List of Abbreviations

ACE	Aids to Communication in Education
ACS	alternative curriculum strategies
APU	Assessment of Performance Unit
ASE	Association for Science Education
ASEP	Australian Science Education Project
AVA	audio-visual aid
BSCS	Biological Sciences Curriculum Study
CAL	computer assisted learning
CDT	craft, design and technology
CLIS	Children's Learning in Science Project
CPVE	Certificate of Pre-Vocational Education
CSE	Certificate of Secondary Education
DARTs	directed activities related to text
DES	Department of Education and Science
GCSE	General Certificate of Secondary Education
GIST	'Girls into Science and Technology'
HMI	HM Inspectorate
LEA	local education authority
MEP	Microelectronics Education Programme
MESU	Microelectronics Education Support Unit
MFY	Mobilisation for Youth
NEA	Northern Examining Association
PSSC	Physics Science Study Committee
RLDU	Resources for Learning Development Unit
SCISP	Schools Council Integrated Science Project
SEC	Secondary Examinations Council
SEMERCs	Special Education Microelectronic Resources Centres
SMA	Science Masters Association
SSCR	Secondary Science Curriculum Review
SSR	School Science Review
TAPS	Techniques for the Assessment of Practical Skills
TIPS	Test of Integrated Process Skills
TVEI	Technical and Vocational Education Initiative

PART I
Science for All

Introduction

'Science for all' has become the slogan of the 1980s for science educators around the world. In this country its origin can be traced to James Callaghan's speech at Ruskin College, Oxford, in October 1976. This speech has been followed by a period of extended debate on the future of the education system, punctuated by a host of official publications. The principle of secondary school science education for all children was clearly established in Curriculum 11–16 (DES, 1977).

> Science education is for all – not just only for those who have the potential to become scientists, technologists or technicians. All have a right to understand and to become involved in problem-solving processes which they will face in day-to-day living and which require the knowledge and disciplines of science ... A science course, therefore, is an essential component of the curriculum of every boy and girl up to the end of compulsory schooling.

A later document, *Science Education in Schools* (DES, 1982), extended this principle to include the primary school by recommending that science education be regarded as a continuum from 5 to 16, and in the most recent document, *Science 5–16* (DES, 1985a), the intention of establishing a firm policy of science education for all children throughout the period of compulsory schooling is made abundantly clear.

> Science should have a place in the education of all pupils of compulsory school age, whether or not they are likely to go on to follow a career in science or technology. All pupils should be properly introduced to science in the primary school, and all pupils should continue to study a broad science programme, well suited to their abilities and aptitudes, throughout the first five years of secondary education.

Throughout this period of debate there have been several other national and local initiatives for science curriculum reform. For example, the Secondary Examinations Council has developed grade criteria for the new General Certificate of Secondary Education (GCSE) examinations; the Royal Society has published two major reports (*Science Education 11–18*, 1982, and *The Public Understanding of Science*, 1985); there has been generous financial support to local education authorities (LEAs) for in-service courses for heads of secondary school science departments and for science coordinators in primary schools; there has been the development work of the Technical and Vocational Education Initiative (TVEI), the Assessment of Performance Unit (APU), and the Microelectronics Education Programme (MEP); a number of LEAs have introduced 'alternative curriculum' strategies (ACS) and a number of institutions have developed new curricula associated with the Certificate of Pre-Vocational Education (CPVE). Each of these developments has implications for the science curriculum. Perhaps the most significant initiative as far as the secondary school science curriculum is concerned is the Secondary Science Curriculum Review (SSCR), established as a five-year development project in 1981 and sponsored by the Association for Science Education (ASE), the Department of Education and Science (DES), the Northern Ireland Council for Educational Development, the School Curriculum Development Committee and the Health Education Council. To an extent, the main orientation of the SSCR proposals has been influenced by two publications from the ASE: *Alternatives for Science Education* (ASE, 1979a) and *Education Through Science* (ASE, 1981). A common theme of these reports and recommendations is the proposal for compulsory science education for all children between the ages of 5 and 16. It is admirably expressed by a Royal Society working party.

> Everyone needs some understanding of science, its accomplishments and its limitations, whether or not they are themselves scientists or engineers. Improving that understanding is not a luxury; it is a vital investment in the future well-being of our society. (Royal Society, 1985)

We concur with these sentiments and maintain that schools should seek to provide, first, a basic science education for everyone and, secondly, a sound basis from which the scientifically gifted can attain the very high standards we can, and should, expect of them. It is our view that we should attain the very highest standards for some and raise the minimum standard for all, and at the same time provide equality of curricular opportunity for all children, regardless of gender, race, social origin or initial aptitude and ability. This curriculum intention has enormous implications for content, methods of teaching and learning, assessment procedures and, as we shall see in later chapters, school organisation.

The notion of 'science for all' is a difficult one; whilst it seems to have gained almost universal support during the past five to ten years (Fensham, 1985), it has gained little in clarity of meaning. It can be interpreted in at least two very different ways. First, that all children will do some science, though that science may differ substantially from school to school, and from individual to individual. Secondly, that the same science curriculum will be provided for all children; in other words, a compulsory, common science curriculum will be provided within and between schools. Those who advocate the former run the risk of differentiating science education into high-status science for some and low-status science for the majority. To an extent, this has been the traditional pattern in the UK, with high-status, academic courses being characterised by their abstractness, competitiveness through individual work (rather than co-operation through group work), and unrelatedness to life out of school. By contrast, low-status non-academic courses are characterised by concreteness of knowledge, less emphasis on written presentation and relevance to everyday life. Michael Young (1976) sees this differentiation as part of a huge conspiracy to maintain a large scientifically illiterate workforce, capable of ready manipulation by politicians. Whilst rejecting Young's assertion that this is a motive for providing differentiated science curricula, we would acknowledge that it is a consequence of so doing. Therefore, we would advocate a common science curriculum, with ambitions considerably beyond those usually associated with courses for the less able, up to a basic minimum level of attainment. Thereafter, we would envisage the provision of science curricula appropriate to the needs, interests, aspirations, and capabilities of the children.

This book is concerned with the nature of the basic, common science curriculum and with the particular provision for those designated as having special needs. At the outset we should also state what this book is not. It is not a course book to be followed. It is not even a set of specific proposals for the science curriculum. Rather, it is an exploration of the issues, as we see them, associated with the teaching of science to children of low educational attainment. From time to time, we feel able to set out principles of procedure and guidelines for action, but nowhere do we feel that we are able to set out the answers to the problems we identify. Whilst acknowledging our shortcomings in this respect, we believe that identification of some of the problems is, in itself, a major step forward in the task of designing more suitable curriculum experiences for these children. More appropriate and better designed curricula will, inevitably, lead to increased personal satisfaction and increased learning. The significance of this order of priority will become apparent later.

Chapter 1 seeks to explore the nature of science education and its place within a compulsory core curriculum. Chapter 2 seeks to identify the 'target group' of children and to locate the primary reasons for their lack of success in the education system, as currently organised. A belief implicit in our thinking is that some 80–90 per cent of 16 year olds can, and should, reach a level of attainment currently regarded as achievable only by the average and above-average child. What is needed for our belief to become reality is a determination that children will succeed, a clear specification of what this minimum level of attainment comprises – perhaps along the lines of the Cockroft Report's 'foundation list' for mathematics (Cockroft, 1982) – and a pedagogy and school organisation that take account of wide individual differences amongst children.

–1
Science education for all

INTRODUCTION

It could be argued that the move to comprehensive education represents a rejection of our historical legacy of differentiated schooling, for it is a rejection of the social class basis of selection suggested by the Taunton Report of 1868 and a rejection of the psychological basis of selection by 'aptitude' suggested by the Norwood Report (Board of Education, 1943). It must be admitted, however, that the trend towards common schooling has not entirely removed selection and that it is a matter of some concern that many comprehensive schools continue to band or stream children on the basis of attainment scores and, thereafter, to provide different curricula for different children. Our views on the appropriateness of ability grouping will become apparent later; our concern at this point centres on the provision of very significantly different curricula. Schools claim to be concerned to assist every individual to achieve her or his full potential, but by providing a differentiated curriculum they ensure that the claim remains a myth and that the opportunities for those labelled 'less able' are severely limited by denying them access to certain sorts of knowledge. If we genuinely believe in equality of educational opportunity we must provide a common, basic curriculum as a logical extension of the comprehensive movement. Those who advocate a non-intellectual or a different curriculum for a group of children identified as 'less able' should be required to justify their stance. We do not need to ask 'Why give children the same curriculum?' Rather, the onus is on the would-be differentiators, and we should ask, 'What justification is there for providing different curricula for different children?'

The Newsom Report (DES, 1963) recommended alternative, skills-based and practically oriented courses for less able children ('young school leavers', as the report calls them) because they would use those skills in adult life. Since the 'young school leaver' was to be identified after only two years in secondary school, it is abundantly clear that such a restricted curriculum would enable the child to do little other than menial or manual work. Failing to provide children with higher-level knowledge and skills denies them the freedom of opportunity we profess to espouse.

An argument for curriculum differentiation on the grounds of cultural differences has been advanced by Bantock (1971). He claims that a high-level curriculum oriented towards 'mind knowledge' is appropriate for the elite minority, who exhibit 'heightened consciousness, a high degree of literacy, intellectual interests and spiritual consciousness', and that a non-literary curriculum is more appropriate for the mass of the people, who are characterised by 'direct sensuous awareness ... and a certain narrowness'. This line of argument, which seems to derive directly from nineteenth-century 'mentality theory' (Shapin and Barnes, 1976), is curiously out of touch with contemporary concerns with equality of opportunity, freedom of choice, and social justice. It would do little to ameliorate the unfortunate consequences of Britain's rigidly class-stratified society.

A third justification of curriculum differentiation can be found in the rhetoric surrounding many contemporary proposals for an alternative, urban-oriented curriculum for inner-city children. This is the kind of argument advanced some years ago by Eric Midwinter (1972) to justify a community-oriented curriculum for Liverpool primary schools. The underlying assumption is that children are better motivated and learn more successfully when the curriculum focusses on their immediate environment and community. As will become apparent later, we put a high curriculum priority on motivation and believe that it is the key to successful learning. We also place a lot of emphasis on relating the curriculum experiences to the children's 'everyday world'. However, we share the views of Merson and Campbell (1974) that community-oriented education for inner-city children is based on a misconception of the nature of contemporary urban life.

> Such areas are characterised by social fragmentation ... cultural pluralism, groups with widely differing and potentially conflicting values ... competition for the possession of limited basic amenities rather than cooperation in their use, to such an extent as to make it unlikely that the term community can realistically be applied to them in any sense that is accepted.

Even if a 'community' were identifiable, a community education would be undesirable. It is our view that a curriculum focussing on the inner city, 'in all its moods and manners, warts and all' (as Midwinter advocates), would represent an abandonment of the goal of equality of educational opportunity in favour of a 'ghetto education' that would serve only the ghetto and would effectively prevent inner-city children from entering the mainstream of social and political life, by denying them access to significant knowledge. It is a certain route to social conflict in our cities. A major task of education should be to ensure that disadvantaged children are provided with more effective means of attaining high-level educational goals, the goals more easily within the grasp of others. Community-oriented curricula confirm and extend disadvantages, rather than compensate for them.

The purpose of this very rapid identification and rejection of three of the several arguments for a differentiated curriculum is to establish a basic principle underpinning this book: that no child should be provided with an alternative, inferior curriculum on the grounds of perceived differences from other children, no matter whether those differences relate to race, gender, social origins, or intellectual attainment. We believe that the arguments against a differentiated curriculum and in favour of a common curriculum are overwhelming – on educational, political, economic, social, and psychological grounds. A socially just democratic society requires equality of educational opportunity. Comprehensive education is a step in the right direction; common curriculum provision is the logical and necessary next step. Society benefits when the skills and talents of all children are fully developed. The divided curriculum, by providing significantly different linguistic and cultural experiences within school, helps to perpetuate the stratified society, encouraging a dangerous polarisation into 'them and us', bosses and workers, leaders and led. At work, at home and in the community, a person's self-respect depends on feeling valued by others. The inevitable consequence of following a second-rate curriculum is a feeling of being second rate – a feeling of inadequacy, frustration at chances missed, resentment at chances never offered, and not being respected by peers and by adults.

At a time of growing social unrest in our heavily populated urban areas, we need to grasp the nettle of equality of opportunity. The first step in that direction is the provision of a common curriculum. If this point is accepted, it then becomes a matter of selecting appropriate content for this common curriculum and finding ways for the less able to learn, experience and benefit from what, until now, has been regarded as the privilege of the elite few and of opening up to those few the worthwhile 'practically oriented' educational experiences traditionally reserved for the less able. In summary, we are arguing that our present practice of providing a differentiated curriculum is inadequate for all children. It deprives large numbers of children designated 'less able' of access to significant areas of knowledge and understanding, and it deprives those children identified as 'academic' of the opportunity of following a curriculum firmly anchored to the real world and to the society they will grow up into. In turning our attention to the science curriculum, we will argue that contemporary science curricula are also failing all children. Our dissatisfaction centres on content, on the implicit message about the nature of science and scientific activity, on teaching/learning methods, and on assessment and evaluation strategies and procedures. Each of these is discussed at some length in later chapters.

Lest all the foregoing be dismissed as hopelessly idealistic in the face of very real differences in aptitude and ability, and the very marked barriers to effective learning experienced by some children, we hasten to say what, in our view, a common science curriculum is not. It is not a set of common learning experiences; it is not identical content; it is not common expectations of eventual attainments and capabilities. Rather, it is a set of common goals –common in so far as they represent a common level of attainment in the basic science curriculum provision; common in so far as they represent meaningful experience of science and scientific activity for all children; common in so far as they embody a determination that all children will be brought to a point at which we can claim them to be scientifically literate.

It is implicit in our proposals that we plan the science curriculum from the 'bottom upwards', rather than the 'top downwards'. What is of most importance, and what is central to our major curriculum goals, must be taught to and experienced by everyone. Certain groups of children will reach levels of attainment well beyond these basic levels. That is to be welcomed and celebrated. Our immediate concern, however, is with the nature of the basic curriculum for the achievement of scientific literacy. What is abundantly clear is that the attainment of a common goal of scientific literacy for a group of children with very different initial experiences and very different capabilities will not be easy. It will require a variety of learning routes and learning methods: not all children learn successfully by the same experiences. It will require the school administration to accommodate rather different demands for staffing, resourcing and time allocation than under present conditions. But if the goal is worthwhile, the cost and effort will be worthwhile. The notion of mastery learning implied in this goal of scientific literacy carries with it the need for a clear specification of minimum levels of expected attainment in the basic science curriculum, the principle of criterion-referenced assessment – perhaps on a test-when-ready principle – and the need for effective and comprehensive record keeping and reporting. These are matters that will be discussed at length in chapter 7.

SEEKING AN ORIENTATION FOR A COMMON SCIENCE CURRICULUM

Curriculum theorists tend to classify approaches to curriculum design as knowledge centred, child centred or society oriented. A properly constructed knowledge-centred curriculum is underpinned by a theory of knowledge. Typical of such theories is Paul Hirst's (1965) classification of knowledge into seven forms of knowledge on the basis of four criteria of demarcation:

1. the characteristic basic concepts,
2. the characteristic structures by which these concepts are related,
3. the characteristic ways in which knowledge statements are tested,
4. the characteristic techniques and skills for exploring experience and generating new knowledge.

Hirst's forms include mathematics and formal logic, the physical sciences, the human sciences, moral understanding, religious knowledge, philosophy, and the fine arts, and his principal educational goal ('the development of rational minds') will be attained, he argues, by a curriculum designed to emphasise the demarcation criteria. Whilst we acknowledge the elegance of Hirst's argument and applaud the curriculum proposals that emphasise disciplinary structure (Bruner, 1960; Schwab, 1964; Stenhouse, 1975), we feel that structure is insufficient in itself to provide an adequate set of principles to guide the development of a common curriculum for scientific literacy. The pursuit of a knowledge-oriented curriculum neglects the human and the personal element and, to an extent, ignores the crucial role of the learner. Science, scientific knowledge and the scientific practice are human constructs, and any curriculum that ignores this fact is, in our view, deficient. It is this personal element in science education, discussed at greater length in chapter 2, that is the key to the educational philosophy underpinning this book.

Thus, our curriculum design is firmly rooted in the child- or learner-centred tradition. However, we would also claim some adherence to theories of society-oriented curriculum design, such as those advanced by Lawton (1983) and Skilbeck (1982). Any proposals for a curriculum for scientific literacy must include a consideration of the impact of science and technology on society and the influence of society on science, scientific research and scientific development. If we were to identify a single goal for our curriculum, it would be that each child would be equipped to be an active participant in a democratic society. There are few, if any, public issues that do not have a scientific or technological dimension. What we seek is an informed and thinking citizenry, capable of considering scientific and technological matters, together with economic constraints, environmental issues, ethical concerns, social and aesthetic considerations. The science curriculum is a major vehicle for the achievement of the breadth of perspective on which responsible decision making depends. Similar views are dealt with at some length in recent work by Aikenhead (1985). Contemporary science curricula in the UK and overseas do not achieve this breadth and cannot claim to achieve scientific literacy. Academic courses often fail to address any concerns other than abstract theoretical ones and, in common with non-academic courses, fail to provide children with adequate insight into the nature of science and scientific activity.

The late 1950s marked the end of a long period of stability in the science curriculum. The publication of the Science Masters Association (SMA) policy statement *Science in General Education* (SMA, 1957) in the UK and the Rockefeller Report (1958) in the USA to a large extent initiated the rapid curriculum development of the 1960s and 1970s – the Nuffield and Schools Council courses in the UK; the Physics Science Study Committee (PSSC); the Harvard Project Physics; the Biological Sciences Curriculum Study (BSCS); the Chemical Bond Approach (CBA); the Chemistry – an experimental science (CHEM) etc. in the USA; and the Australian Science Education Project (ASEP) in Australia. Central to the British initiatives was a belief that considerations of the structure of science, scientific knowledge and scientific method should be principal factors in curriculum design. In other words, they adopted a largely Hirstian approach. There was a significant shift of emphasis towards giving children experience of what scientists do. 'Being a scientist for a day' became a catchphrase and there was an associated shift of emphasis away from book learning towards bench learning.

In many ways, however, these courses have failed to meet the high expectations we had of them. They have not, in general, led to increased interest in science; they have not produced the expected cognitive gains; they have not led to increased understanding about the nature of science and scientific inquiry. The causes of this failure are complex and many faceted, but two major factors can be identified: they were based on an inadequate model of science, and they were based on the mistaken assumption that science and scientific method are best learned by discovery methods. Thus, they were both philosophically and psychologically unsound (Hodson, 1985a, 1986a).

It would, of course, be a gross oversimplification to assume that the emphasis on 'being a scientist' was the only curriculum initiative during this period. Indeed, Roberts (1982) identifies seven different curriculum emphases, which he calls 'everyday coping', 'structure of science', 'science, technology and decisions', 'scientific skill development', 'correct explanations', 'self as explainer' and 'solid foundations'. Clearly, none of these is an exclusively correct view, and shifts of emphasis are apparent with time. Fashions come and go. At present there would seem to be an increasing move away from education in science and towards education about science. Concern with social, political and technological issues is apparent in recent proposals from the ASE (1984) and the SSCR (1984). What is distressing, from the learners' point of view, is that reform movements tend to lose sight of other emphases in their pursuit of their particular orientation. Social, moral and economic issues are important, of course, but so too are considerations centring on the structure of the discipline. We need to include value issues as well as, not instead of, concern with the products and processes of science.

The two HM Inspectorate (HMI) surveys of education published in the late 1970s express concern about much of what goes on in schools under the name of science education. For example, in nearly half the schools visited, HMI saw 'serious cause for concern about the standards achieved by pupils'. They go on to say that:

> In roughly 10% of the schools, some of the pupils in the fourth and fifth years were provided with a balanced science curriculum.... No school was found ... which provided balanced science courses for all pupils up to the age of 16 plus.... Examination success ... in a number of schools ... was regarded by teachers as being incompatible with a teaching style which sought to develop analytical and predictive thinking.... In about one third of all the schools the teaching of science was always or nearly always overdirected, with insufficient pupil activity. (HMI, 1979)

The primary school survey makes similarly depressing reading for science educators.

> Few primary schools visited ... had effective programmes for the teaching of science ... the progress of science teaching in primary schools has been disappointing; the ideas and materials produced by curriculum development projects have had little impact in the majority of schools.... Many existing teachers lack a working knowledge of elementary science appropriate to children of this age. This results in some teachers being so short of confidence in their own abilities that they make no attempt to include science in the

curriculum. In other cases, teachers make this attempt but the work which results is superficial. (HMI, 1978)

There is special concern about the science education provided for the children who are the subject of this book. For example, paragraph 86 of *Science 5–16* (DES, 1985a) states that many pupils with special educational needs are not offered 'as much science, or science in as much depth and breadth, as is desirable', and Brennan (1979) has found that 'profitable work' in science occurs in only one in nine special schools. HMI (1984) report that many schools find it difficult to meet individual needs and that the aims, priorities, management and teaching styles adopted often actively work against success for slow-learning and academically less successful children.

Whilst accepting much of this criticism, we feel that these reports do not identify the very major advances that have been made in science education during the past two decades. There has been a significant and laudable move towards more carefully structured courses and more active involvement of pupils. Many schools have produced extensive and elaborate resource materials for use with less able children who attain learning standards considerably in excess of what was regarded as possible only ten years ago. New assessment techniques have been introduced and are now widely practised; even a casual reading of test materials produced in the early 1960s will reveal the extent of progress in this area. All this has been achieved through the time, energy, expertise and commitment of many dedicated teachers, and to dismiss this progress by wholesale condemnation of existing provision is ill-informed, discourteous and likely to be counter-productive.

This is not to say that 'everything in the garden is rosy' and that all is a matter for rejoicing. Far from it. There are many aspects of contemporary science curricula where we should look for improvement:

- science begins too late and, for many, ends to soon
- many courses have too much content
- many courses have content that is too difficult and too abstract
- many courses make little, if any, attempt to deal with philosophical, historical, social, economic, moral and ethical issues
- there is too little integration of the sciences and of the sciences with other disciplines
- too little attention is paid to individuality of learning needs and individuality of response; teachers are too 'group conscious', rather than 'individual conscious'
- too little care is taken to ensure continuity between the primary and secondary sectors.

As a consequence of these and other deficiencies to be identified later, large numbers of children opt out of science at the earliest opportunity and large numbers of children leave school ignorant of, or even hostile to, science. The Royal Society points out that hostility and indifference to science and technology, whether by shopfloor workers, managers or investors, weakens British industry. There would be considerable competitive advantage, they argue, if 'those who hold positions of responsibility had at least some understanding of what science and technology can and cannot achieve, and were better able to call for and evaluate advice on scientific and technological issues' (Royal Society, 1985). HMI (1979) report that one-eighth of secondary school children study no science after year three. More significantly, 55 per cent study only one science subject in years four and five.

There are, of course, children who study too much science in school – some seven per cent follow three science subjects to O level (HMI, 1979), with the consequent severe restrictions placed on the rest of their curricular experience. Most importantly, contemporary science curricula neglect the socio-economic and moral-ethical issues in science. What does the future hold if we produce trained scientists and technologists who are insensitive, ignorant of wider issues, and incapable of using their knowledge wisely? This state of affairs should not be allowed to persist. Not only is it detrimental to the supply of scientists and technologists, on whom our material well-being depends, but it runs counter to the demand for an educated citizenry capable of active participation in a democratic society. We believe that education for leisure is as much a function of education in science as it is of education in the arts. We would argue that all children have the democratic right of access to the kind of science that enhances their understanding of the natural world, provides access to leisure pursuits with a scientific or technological basis, such as photography or computing, and assists them towards an appreciation of the issues raised in a political debate or dealt with in a television programme.

If we are to enter the last decade of this century with any degree of economic and social confidence in Britain's future, curriculum designers and teachers must engage in an overall reconstruction of the science curriculum. Our aim should be a balanced science curriculum for everyone. The authors of *Science 5–16* admit that much pruning of the factual and theoretical parts of the existing curriculum will be necessary if this goal is to be achieved, and that, for many, there will need to be a much more generous allocation of curriculum time for science. They recommend 10 per cent in years one and two, rising to 15 per cent in year three and to 20 per cent in years four and five. Whilst the document does not expressly recommend integrated science, it does say that 'each pupil is entitled to a programme of study which … will incorporate substantial elements from each of the three main sciences'. The policy statement goes on to identify ten basic principles underpinning good science curriculum provision: breadth; balance; relevance; differentiation (so that pupils of all abilities are catered for); equal opportunities for girls and boys; continuity between schools; progression (by making science a coherent series of experiences from age 5 to 16); links across the curriculum (especially with craft, design and technology (CDT), maths and language); teaching methods with a practical and problem-solving approach; sound assessment methods.

The proposals we wish to advance in this book – an initial course in integrated science which is largely process based, leading to a consideration of separate sciences later on – are broadly in line with the pronouncements in *Science 5–16* and with the government policies for education in the twenty-first century, published under the title *Better Schools* (DES, 1985b). However, they do have significant implications for resources, staffing levels and school organisation (see chapter 8). It is a matter for regret that *Science 5–16* fails to recognise the very considerable implications for resources and training of their far-reaching proposals for curriculum renewal.

The foregoing considerations, especially the requirement to meet the three major curriculum orientations (the nature of knowledge, the needs and aspirations of individual learners, societal issues), lead us to propose that a basic education aiming at scientific literacy should contain (at least) the following:

- science knowledge – certain facts, concepts and theories
- applications of knowledge – the direct use of scientific knowledge in real and simulated situations
- skills and tactics of science – familiarity with the processes of science (classifying, hypothesising, etc.) and in the use of apparatus and instruments
- problem solving and investigations – application of knowledge, understanding, skills and tactics to real investigations
- interaction with technology – practical problem solving, emphasising scientific, aesthetic, economic, social, and utilitarian aspects of possible solutions
- socio-economic-political and ethical-moral issues in science and technology
- history and development of science and technology
- study of science and scientific practice – philosophical and sociological considerations centring on scientific methods, the

role and status of scientific theory, and the activities of the community of scientists.

Such a curriculum – details of which will be discussed in later chapters – would go some way towards achieving our goal of universal scientific literacy.

Closely related to the notion of scientific literacy are the concepts of technological literacy and computer literacy. We envisage the curriculum taking account of both these perspectives. As far as technology education is concerned, we recognise that many scientific activities involve technology of various kinds and that technological problem solving involves scientific ideas and theories, so that science and technology are very closely related and, sometimes, indistinguishable. However, in this book we make no pretence to discuss technology education, as such, beyond the suggestion that it should focus on the following:

- the mechanisms that underlie the working of the technology under consideration
- the techniques required to make use of the technology
- the functions that the technology is capable of carrying out
- the applications of the technology to real problems
- the social context and implications of the development and application of the technology.

These aspects, together with the practical problem-solving activity referred to earlier under the heading 'Interaction with technology', would constitute a programme capable of ensuring universal technological literacy.

As far as computer literacy is concerned, there is general agreement that its achievement is 'a good thing', but there is little general agreement about what it is. An inspection of the literature reveals at least three possible orientations:

1. mastery of technique
2. access to and ability to use computers as tools
3. awareness of and understanding of computers in their various socio-economic contexts.

Traditionally, computer studies courses have emphasised mastery of technique – learning how a computer works and acquiring expertise in simple programming techniques. Indeed, Luehrmann (1981) declares that 80 per cent of a computer literacy course should comprise these kinds of activities. We strongly oppose this highly restricted and introverted view of the nature of computer literacy on the grounds that it ignores the wide issues concerning the applications and implications of computer technology. We would also argue that it is based on the unjustified assumption that computer literacy requires technological knowledge when, in reality, it is perfectly possible to program without having knowledge of how computers function and perfectly possible to use a computer, for a variety of purposes, without being able to program. Given the rapid pace of computer technology development, there is also a very real danger that courses emphasising technical matters are permanently out of date. Knowing that programming is necessary, and that a computer can do only what it is programmed to do, is an essential part of basic computer literacy. And, we admit, the best way of learning this fundamental principle might be through engaging in simple programming exercises. But this is entirely different from orienting the course towards the acquisition of programming techniques.

The second orientation is on making use of the computer as a means of communication, as a means of handling data and solving problems, and as a device for learning. The emphasis here is on familiarity with the computer as a tool, on recognising situations when the use of a computer is appropriate, on selecting and implementing a computer solution. However, arguing that, because information is increasingly stored and handled by computer, it is necessary to provide a separate course dealing with the computer is as absurd as the notion that we should provide a course in books, because many powerful and interesting ideas are recorded in book form. Just as books are used across the curriculum, so computers should be used across the curriculum. For example, experience of data handling might be most appropriately and most frequently located in history or in science, word processing in English and simple programming techniques in mathematics. In other words, we are arguing that children should acquire computer literacy in familiar learning environments. Using the computer in other classroom activities emphasises that it is a tool, rather than an object of study in itself. In this way, we maintain, we can de-mystify the computer. We shall argue repeatedly throughout this book that a major task confronting the teacher of underachieving children is building up the learner's confidence and self-esteem. Refusing to afford computers a special place in the curriculum is an important first step in building up children's confidence in their ability to use them. We nevertheless recognise that, in the same way that it is common practice to try to ensure that all children are equipped with the basic reading and study skills necessary for the productive use of books, it will be necessary to provide all children with the basic keyboard skills required to make productive use of the computer. This 'new literacy', as we might term it, includes the ability to write at the keyboard, to read a VDU and to interrogate an information retrieval and manipulation system. But, we maintain, such skills training should be part of a basic study skills course. It may also be worth noting at this point that there is a strong possibility that, as computers become increasingly 'user friendly', the need for extensive study skills courses will decline.

The third orientation for a computer literacy course is awareness of computer technology in its socio-economic context. This is quite distinct from either of the other two approaches, in that it confronts the implications of computer technology for consumers, workers and citizens. In other words, the focus of attention is the interaction of computer technology with society. Clearly, that interaction is a consequence of both the capability of the technology and the human decisions about its deployment. Thus, we would argue for a two-pronged assault on these issues – first, through the technology curriculum and, secondly, through the humanities curriculum. We see the computer as the subject of study, alongside other contemporary technologies, within a much expanded technology curriculum, organised on the basis of the five aspects outlined above (p. 17). During the course, the emphasis would usually shift from a technical to a functional perspective and, finally, to a cultural perspective. It is at this point in the curriculum that the humanities would have an important role in dealing with social, economic, political and ethical issues, and with the development of attitudes and interests. Where better to deal with the storage of and access to information, and their social and political implications, than in a history or literature course? Indeed we would argue very strongly for a much closer curriculum relationship between science/technology and the humanities. On the one hand, issues concerning the human application and direction of science and technology can be dealt with through literature and history and, on the other hand, the science/technology curriculum can be a useful vehicle for the enhancement of literacy. The precise relationship we envisage will become more apparent in later chapters.

Thus far, we have argued for the dismantling of computer literacy courses and the dispersal of their various elements to other areas of the curriculum. However, we admit that there are powerful arguments in favour of separate courses in computer literacy: such an approach localises the problems; it makes optimum use of the (few) staff with computer skills; it is based on the departmental structure that underpins most curriculum practice; it is 'high profile' and lets the community know that schools are 'doing something' about the 'micro-revolution'. But are such courses of any educational value in their own right? We would argue that they have only limited value and that computer literacy is best achieved by a combination of the following:

- a short study skills course for equipping each child with basic keyboard skills
- dispersal of the 'tool using' aspects to appropriate parts of the curriculum
- awareness of technological function, social implications, and so on, to be located within the technology curriculum, in accordance with the outline described above, and within the history, literature and social studies curriculum.

In essence, what we are saying is that computer literacy should emerge from the total curricular experience of the individual child, through contact with the computer in as many contexts and educational roles as possible. We believe that such an approach is desirable for all children, but that it is crucial for slow learners, who are likely to be easily disheartened by an approach that places heavy emphasis on technical matters. We see each subject in the curriculum having a role in bringing about universal computer literacy.

This is an appropriate point, then, at which to emphasise a fundamental principle underpinning our views about science education. Unlike many of our contemporaries, who see science education solely in scientific terms, we regard the science curriculum as a vehicle for the attainment of other educational goals. In addition to its role in achieving scientific literacy, we see it as a means of contributing to the attainment of computer literacy, we see it as a vehicle for enhancing numeracy, through its concern for such things as measurement, numerical data manipulation and graphical representation, and we see it as a vehicle for enhancing literacy, through the provision of reading and writing tasks of various kinds. But above all, for the children who are our principal concern in this book, we see it as a vehicle for the development of personal and social skills and for the enhancement of self-esteem. Thus, we see the science curriculum as having a much wider responsibility than has been traditional in secondary schools – wider in the sense of encompassing a broader range of goals; wider in the sense of concern for all children, rather than concentration on the academic elite. In later chapters we hope to show how the science curriculum can, and should, be adapted to meet these wider responsibilities.

At this point we should remind ourselves that a common science curriculum has to cater for two broad groups of pupils: those who will study science at a more advanced level, and may even go on to engage in science-related employment, and those who will not. Thus, the science curriculum must be a sound and adequate preparation for later study and must ensure scientific literacy for those others – the majority – who will opt for alternative pursuits. In meeting these two needs it must teach science and it must teach about science. And it must make science accessible to all children, including those who traditionally have followed no science course or, at best, a rather low-level science programme. Ensuring accessibility to scientific literacy is the general problem confronting us in designing a common science curriculum. Accessibility can only be improved by learning more about the learner. In chapter 2 an attempt is made to identify some of the factors that constitute barriers to effective learning, contribute to underachievement in science, and thereby deny access to science to large numbers of children.

2

Identifying the scientifically less able child

THE TARGET POPULATION

About 25 years ago a young science graduate began a summer vacation job in a large Somerset bakery, in order to earn some money prior to starting an initial teacher training course. Although the pay was quite good, the work was physically demanding. The factory was large and badly ventilated, which made the summer afternoons almost unbearably hot. One of the jobs the student had to do involved dressing in protective clothing and collecting trays of freshly cooked pies as they emerged from a long open oven. Constant activity was necessary to keep up with the moving oven belt, otherwise the trays of scalding fruit mixtures spilled onto the factory floor. There were many other, equally unpleasant, tiring and repetitive tasks that had to be performed. The workforce consisted largely of strong young men between the ages of 15 and 25 who often worked long hours of overtime to boost their weekly wage packet. These young men were the 11+ failures of those days who had followed courses in the secondary modern schools of the late 1950s, and who had failed to qualify for the many semi-skilled jobs and apprenticeships then available. By whatever criteria might be applied – of parental background, of their own personal prestige and standing in the larger community, or of their attitude to life – they were unmistakably 'working-class' people. It is easy to imagine that in today's comprehensive schools, in the current climate of youth unemployment, and in the inner cities they would fit fairly neatly into the category of children that educationalists call the 'underachievers'. But the question is, in spite of their underachievement, were they also the 'less able'? In other words, if they could have been motivated to respond to schooling, would they have been able to take greater advantage of it in terms of employment?

Relief from the drudgery of the work in the bakery was obtained during breaks in the locker room. Here the men drank tea, smoked cigarettes and talked about life outside the factory gates. One of the constantly recurring themes of this conversation was horses. To hear those people talking about horses was an education in itself; they became suddenly articulate, well-informed and thinking practitioners of a complex activity. The speed with which they could calculate the odds on the tote or an accumulator was phenomenal. And the permutations could be worked out at a moment's notice as fresh information was gleaned about the state of a particular race course, a trainer's pending bankruptcy, the type of shoe worn by a fancied horse on its last appearance, or rumours about a jockey's weight problem. These young men suddenly appeared as agile mental mathematicians; they had a grasp of specialist language that encompassed an entire jargon; they understood elements of the science of soils and weather and the geography of racecourses; they could account for independent variables in an equation with the apparent facility of a research scientist; their powers of observation appeared to be as acute as many a trained biologist. In this context it was the student who perceived himself as 'less able', who lacked the self-confidence to enter into conversation on their territory, and who became less motivated to compete with the acknowledged experts in the field as the weeks went by.

In itself the term 'special needs' carries no educational meaning. Just as any group of two or more children must be considered as a 'mixed ability' group, so every child and every group of children has its own 'special needs'. Clearly, however, the term is meant to imply more than just 'all children are special'. Any assumptions that are made about what the phrase 'special needs' might mean in terms of science education will control to a greater or lesser extent not only the specific content of the remainder of this book, but also many of the implicit or hidden messages that it carries. It is, therefore, important for readers to be aware of precisely how 'special needs' are interpreted in the book, in order that they can relate its contents, suggestions and implications to specific school situations. It is equally important to understand that certain aspects of what have been traditionally viewed as 'special needs' are specifically excluded from our overview.

Dyslexic children have special needs, and so do children who are physically disabled and studying science (Jones, 1983; Ayres & Hinton, 1985). Children with moderate and severe learning difficulties also clearly have special needs (Rose, 1981). Then again, there is that group of children designated 'gifted', for whom special schools have been created in the UK (Chetham's School of Music and the Royal Ballet School, for example), in America and behind the Iron Curtain (Freeman, 1985). None of these categories is in itself very large. Indeed, prior to the Education Act of 1981 (DES, 1981b), children with 'special educational needs', comprising all but the gifted group, were seen as constituting only about 2 per cent of the school population. It was Section 1 of this 1981 Act that extended the concept of special needs to the extent of establishing that:

> a child has special educational needs if he has a learning difficulty which requires special educational provision to be made to meet these needs. 'Learning difficulty' is defined to include not only mental and physical disabilities, but any kind of learning difficulty experienced by a child provided that it is significantly greater than that of the majority of children of the same age.

Science teachers reading this might be excused for assuming that the phrase 'significantly greater' carries with it connotations of some quantifiably measurable component, above or below which the child automatically becomes in need of special provision. That is not the case however; indeed the concept of 'special needs' is so vague in this part of the Act as to be virtually meaningless as a guide in any practical teaching situation. The fact is that the definition of 'special needs' in the Act is based largely upon the Warnock Report of 1978 (DES, 1978a), which indicated that up to 20 per cent (the famous 'one-in-five') of all school pupils may at one time or another in their school careers require some form of special educational provision, in science as much as in any other area of the curriculum. Indeed, the science teacher has good reason to see the problem as involving potentially much greater numbers than the original Warnock 'one-in-five' for at least two reasons.

In the first place, science is seen as a difficult subject for children in general (Forrest *et al.*, 1970), and for girls in particular (Kelly, 1981a). Indeed, the Department of Education and Science itself, in a document entitled *Science Education in Schools: a consultative paper* (DES, 1982), openly admits that 'science is a taxing subject', yet nevertheless argues that it should be taught to all pupils in secondary schools, and 'not only to those who have the potential to become scientists' (DES, 1977, 1981a). Thus the challenge for the science teacher in ordinary schools is established. The current concern that this difficult subject is taught to all pupils is heightened by the deliberate inclusion of children who have learning difficulties that are 'significantly greater' than those of the majority of the school population. It is reasonable to conclude that, because of the inherently difficult nature of the subject, there will be more children displaying such 'significantly greater' problems than in some other curriculum areas.

In the second place, Warnock's 'one-in-five' is almost certainly a conservative estimate of children likely to have special learning difficulties in certain types of school. This is due not only to the difficulty of the subject itself, but also to the pattern of distribution of such children in the country's schools. Epidemiological studies indicate a far higher incidence of many educational, social and psychiatric problems in children living in large cities, and especially those in socially deprived areas. HMI Fish (1982), for example, makes the point that in such schools up to 40 per cent of the children could be described as of 'below average and less successful', and those of us working in such inner-city conurbations as Manchester, Salford and Stockport would concur with his observation.

This, in essence, comprises the target group of our discussion: up to 40 per cent of any school population. This 40 per cent represents the majority of those children who traditionally have not been considered worthy of public examination at 16+; the group unlikely to obtain a minimal Grade 5 at CSE level. It is the group identified by HMI during their visits to secondary schools between 1980 and 1982 as 'slow learning and less successful pupils', that is, 'those whose general ability and rate of progress are significantly below average, and those who have some specific learning problem which leads to a general or specific lack of success' (DES, 1984).

SOME CHARACTERISTICS OF UNDERACHIEVERS IN SCIENCE

This DES publication provides us with eight initial clues to some of the underlying reasons for low attainment and we shall return to these later. Foster (1984) and Jenkins (1973) focus more specifically on some of the characteristics and traits displayed by low achievers in science lessons. Table 2.1 is a composite list representing a synthesis of suggestions. Although this is a long list, we should bear in mind that it is not comprehensive. It does not, for instance, include any reference to home background or ethnic origin. Additionally, we should be careful not to fall into the trap of making unwarranted assumptions on the basis of certain behaviours. Presenting untidy work could, for instance, be taken as evidence of a lack of psychomotor skill. On the other hand, it may be caused by a quite different problem or occur in tandem with a different problem. Take Figure 2.1 as an example. The writing in the first line is perfectly legible, although there are indications that the writer is not finding it easy. One hour later, after some experiments on heart beat, and after only seven short lines of writing, it is almost illegible. But the illegibility cannot be explained solely on the grounds of a lack of psychomotor skill or else the first line would have been equally illegible as the last.

Of course, not all low achievers will display all of the characteristics in Table 2.1 at any one time, nor perhaps the majority of them all of the time. But there can be little doubt that, for whatever

Table 2.1 *Some traits shown by underachievers*

Perform badly in school work
*Say they are bored
*Are restless
 Have difficulty in relating to one lesson to next
 Have limited concentration
 Have poor memories
 Have difficulty in following instructions
*Indulge in idle chatter
*Fail to do homework
*Fail to take care over work
*Rarely have pen, pencils, books, etc.
*Lose things
*Respond better to individual attention
*Disrupt other pupils' work
*Are distrustful of teachers and of authority
*Form unstable or weak friendship bonds
*Are often late for lessons
*Are absent more frequently than other pupils
*Claim that what they learn is of no use
*Feel that school is an imposition
*Wish to leave school to earn money
*Express non-involvement in their form of dress
*Are disrespectful of property
*Are attention seeking
 Break things
 Have difficulty in setting up apparatus
 Present untidy work
*Dress untidily

reason, the low-achieving child could be expected to display consistent behaviour patterns somewhere within this spectrum. An examination of the traits in Table 2.1 reveals evidence of an underlying pattern or consistency. Reference to the list shows that the majority of the behaviours have been marked with an asterisk, and it is our contention that these marked behaviours display an important feature in common, and just as significantly a feature not in common with the unmarked behaviours. The unmarked behaviours could be argued to have cognitive or psychomotor origins. In particular, 'limited concentration' and 'poor memories' may be cited as evidence of limited cerebral agility or cognitive skills and, as we have argued 'difficulty in setting up apparatus' could be taken as evidence of limitations in psychomotor skills. Common sense would dictate that progress in science by children with limitations in both the cognitive and psychomotor domains would indeed show itself in a general lack of ability in the subject.

There can be no gainsaying the general proposition that some children are significantly lower achievers in science simply because they are less able than their peers, and this is reflected also in the list

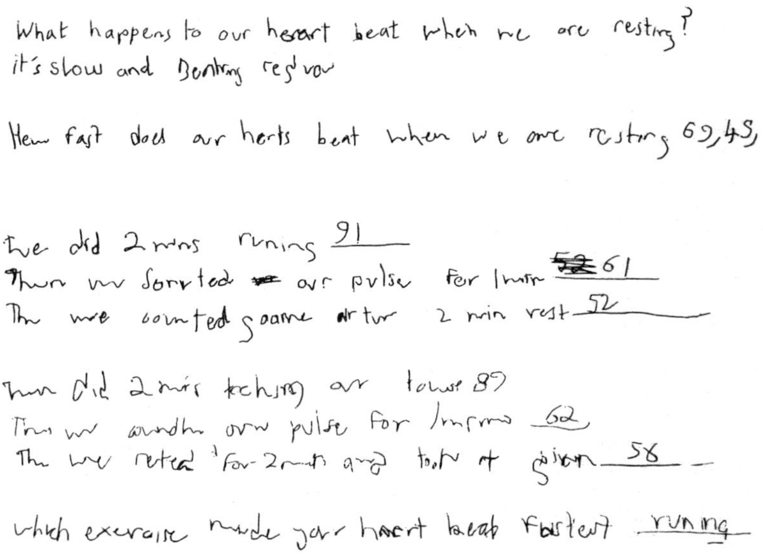

What happens to our heart beat when we are resting?
it's slow and bumping resrvor

How fast does our herts beat when we are resting 69,49,

I we did 2 mins runing 91
Then we sorvted our pulse for 1min 61
The were counted some artur 2 min rest 52

Then did 2 mins kching our tower 89
Then we counted our pulse for 1minms 62
The we reted for 2mins and toon t given 58

which excerise made your heart beat fastest running

Figure 2.1 Writing typical of an underachieving child

of clues provided by the DES (1984). Three of the eight reasons that HMI identify for low achievement fall into the category of limited psychomotor or cognitive skills. However, the other five do not, and nor do the bulk of the traits in Table 2.1. It is just as logical to interpret the asterisked behaviours in Table 2.1 in terms of a lack of an affective commitment to learning in science as it is to a lack of cognitive or psychomotor skills. Indeed, being 'bored' is precisely that: a condition of apathy. And apathy is a feeling all of us have experienced at one time or another, more often than not brought on by conditions that we do not like being imposed upon us, rather than by any innate inability on our parts to cope or to respond. It is a common phenomenon to hear even (especially?) the cleverest of children complain that schoolwork is, at times, 'boring'.

THE IMPORTANCE OF THE AFFECTIVE DOMAIN

In the mid-1950s, Benjamin Bloom and his co-workers identified a trilogy of domains of educational objectives, which were seen to encompass the whole range of experiences to which children are subjected in school. Bloom called them the cognitive (intellectual) domain, the affective (emotional) domain, and the psychomotor domain (which is concerned with motor skills such as are required for writing, playing games, etc.). The cognitive domain emphasises the objectives of science learning in terms of the type of intellectual gains a child might make, ranging from the simple accumulation of knowledge at one end of the domain, through the more complex intellectual activity involved in manipulating that knowledge successfully so that he may be said to comprehend it. At the top of the hierarchy of cognitive activities is the process of evaluation, seen by Bloom (1956) as the highest of intellectual skills, subsuming as it does elements of analysis and synthesis in the ability shown by, say, a doctor in writing a critical report of the treatment given to a patient. The affective domain is concerned with 'a feeling of tone, an emotion, or a degree of acceptance or rejection', which every learner must necessarily bring as part of his overall attitude to a learning task (Krathwohl, Bloom and Masia, 1964).

Table 2.2 summarises the hierarchy of the affective domain. At the bottom of the scale, we can see that the child is merely aware of the stimuli being received. The teacher's voice is heard droning on as it were, but the child is hardly able to relate what is heard to anything that is in any way meaningful. The science teacher would quickly come to recognise such pupils as uninterested in science. These would be the children who 'fail to do homework' and 'are absent more frequently than other pupils', for example. Were this affective commitment never to increase, these children would very soon become identifiable as significantly slow learners. Hopefully however, the climb up the affective hierarchy of commitment will be stimulated by the kind of science that is presented to the child and more positive attitudes will be soon displayed. The child will, at least on some occasions, show a willingness to attend, even if at times there are still claims to boredom. At a higher level of affective commitment still, the child will begin to respond to science lessons with stronger positive feelings, especially perhaps when, in the terms of Table 2.1, the opportunity is given to respond to individual attention. Eventually, the science teacher is hoping that all children will go out of their way to respond, and this would be exemplified in such behaviours as pupil-initiated questions during the course of a lesson, whether a simple request for help in the measuring of a volume of liquid or hypothesising about the meaning of a set of results obtained during a class experiment. Even at this apparently quite advanced state of affective commitment to science, the child has progressed only a relatively short way up the ladder of commitment displayed by full-time professional scientists, who value science as a pursuit in its own right, and who organise their lifestyles around their need to enhance their personal science value systems. These higher levels described by the affective domain are hardly relevant to the immediate issues at stake in the teaching of science to low achievers.

Table 2.2 *Summary of the affective domain (after Krathwohl, Bloom and Masia, 1964)*

TOP OF HIERARCHY: HIGHLY MOTIVATED CHILD	
Level	*Behaviour exemplifying motivational state*
5.0 CHARACTERISATION	Develops regulations for personal lifestyle, e.g. gives up smoking Judges problems in scientific context, e.g. possible effects of smoking
4.0 ORGANISATION	Develops plans, e.g. for science project Forms judgements about scientific ambiguities, e.g. nuclear energy
3.0 VALUING	Confidence in own powers of reasoning, e.g. will attempt to explain to peers Consistency in response to science, e.g. no longer absents self from lessons
2.0 RESPONDING	Satisfaction in response, e.g. sometimes asks for work in science area Willingness to respond, e.g. will read a book about science, however short Willingness to comply, e.g. with regulations about safety in the laboratory
1.0 RECEIVING	Selected attention, e.g. will negotiate a topic Willingness to receive, e.g. usually turns up to lessons; usually not disruptive Passive awareness, e.g. when turns up, makes little effort to be involved
BOTTOM OF HIERARCHY: DEMOTIVATED CHILD	

It has been assumed in the past that affective gains automatically accrue to a child as a result of the cognitive gains he or she makes. Thus, as happens with many of the more able children who study science and technology in school, as they get to know more science they begin to recognise the emergence of overarching patterns that increase their understanding of the subject. This increased understanding itself serves as a reward for their labours, and induces a greater willingness to respond to the learning process, a greater commitment and an increasingly active interest in what the subject is about. This growing affective response itself provides a feedback effect on the accumulation of more cognitive skills as more learning is stimulated. In effect, however, the original stimulus that has catalysed the chain reaction is taken to be a cognitive stimulus, and this is reflected in the largely cognitive-oriented aims and objectives typified in many of the professionally published curricula over the past two decades or so. It is axiomatic that this is not the pathway by which low achievers in science and technology make headway. If it were, we should not be faced with the problem of low achievement. It is our concern to see a reversal of this priority.

Krathwohl and his co-workers make all-too fleeting reference to the relationship between the cognitive and the affective domains. Their analogy of a curricular aim being seen as a wall to be scaled is worthy of some development. Against the wall rests a ladder, each rung of which, they suggest, represents one of the increasingly complex cognitive activities involved in the learning process (see left-hand side of figure 2.2). The rungs, however, are rather far apart. If we imagine that more able children have slightly longer legs than less able children, it is possible to see them being successful in gaining the next higher rung. For the less able child, however, extra assistance is required. This is available in the form of a parallel ladder resting close to the cognitive ladder. Its rungs, too, are rather far apart, so that success in scaling the wall is also impossible if access is available only to this one ladder. Its rungs represent the various increasing affective commitments required for successful learning. However, if the two ladders are placed side by side, in such a way that the rungs of one ladder fall between the rungs of the

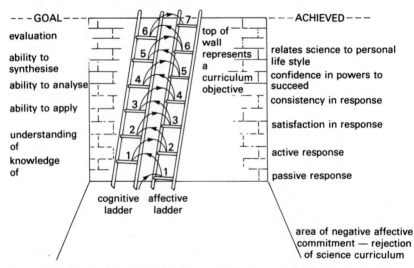

Figure 2.2 Hypothetical relationship between affective and cognitive achievement, stressing importance of the affective domain

adjacent ladder, even children with relatively short legs are able to make some, albeit maybe precarious, progress. This they do by traversing between the two ladders.

Unfortunately, it is not yet possible to identify specific linkages between the two ladders. We cannot say precisely what are the motivating experiences necessary to facilitate access to rung two of the cognitive ladder from rung one. This is mainly because of the individual nature of the problem. For example, and especially in science, girls are likely to be motivated by different experiences from boys. This means that individual teachers of science to children with learning difficulties have to take a personal responsibility for alerting themselves to the needs of individual children. There are guidelines which can be applied, and indeed this book is largely an attempt to investigate these guidelines in some detail. We know that we want to involve children in active learning methods (this does not necessarily mean 'practical work', see chapter 6). We know that science has to be seen by the children as real and relevant. We know that the children must be given tasks at which they can succeed. We know that the children must be encouraged to take more responsibility for their own learning. We know that the key to evoking response lies in the skill of the teacher to communicate with the child. Teachers 'need to see pupils as struggling to impose meaning on their experience and curricula should be planned to enable pupils to consider, contemplate, expand, modify or change their views of the world' (Osborne & Wittrock, 1985). We would add that children with learning difficulties need, above all, to change their views of themselves.

The reader will be aware that the target group of children that forms the subject of this book has been variously described as having 'special needs', being 'less able', having 'learning difficulties', and, latterly, as 'low achievers'. This evolution of the concept of special needs through a concept of less able to a concept of low attainment is crucially important. We do not, of course, deny that there is a substantial proportion of children who are low attainers because they are less able than the norm. What we do refute quite categorically is that low attainment is necessarily, or even usually, a reflection solely of low ability *per se*. In other words, whilst we would not expect less able children to achieve, in relative terms, as well as their more able peers, we would expect them to achieve more than they often do. The story related at the beginning of this chapter is but one in a plethora of anecdotal evidence suggesting that children's performance in school science is at least as much a matter of the 'willingness' of a child to respond as it is of any innate 'ability' to respond. There can hardly be any teacher who does not have his or her own personal experience of the metamorphosis that can occur in children's behaviours as a result of a change of attitude. Not all of the underachievers in science can be motivated. Not all of even the most able children are motivated by the subject. They make a success because, in the terms of figure 2.2, they have long enough legs; that is, even if they do not intend to become scientists or technologists, or even to work in areas remotely concerned with the subjects, they understand that future needs will require evidence of some expertise in the science area. This insight is often unavailable to low achievers, so that without immediate motivation they are unlikely to be prepared to endure the hardships presented by inadequate curricula. Nor are they encouraged to do so when, as is the case in some depressed inner-city areas in the 1980s, 80 per cent of them will be unemployed upon leaving school at the age of 16.

MOTIVATION THROUGH THE CURRICULUM

It is by no means a new idea that the principal challenge to the science teacher is the challenge to motivate the child. It is no coincidence, for example, that the first of the aims listed by the curriculum developers of the Nuffield Biology Course (1966) was to 'develop and encourage an attitude of curiosity and enquiry', which is clearly a long-term affective aim, designed to instil into children a lifelong motivation in the subject. More apposite to less able children and those who are underachieving is the LAMP project (Bowers, 1976), a science course sponsored by the Association for Science Education, and specifically designed for secondary school pupils of low academic motivation. Before listing their aims, the authors attempt to diagnose particular problem areas likely to be encountered by these children: 'In broad terms, the target population was seen as those pupils who for reasons of ability or *lack of motivation* are technically not examinable at C.S.E. level' (our emphasis). And again, under 'issues and problems', the authors of the project state that one of the most common of the problems raised was '*Low motivation frequently found* in 13–16 year old pupils already conditioned to accepting low success in school work' (our emphasis). They go on to summarise the aims that those teachers involved in the project had in mind. There are eight of these, and it is worth listing them.

1. The formation of maturing relationships with adults and peers.
2. The development of self-confidence and esteem through successful experience.

3. The development of skills of communication and number work through practical experiences.
4. The development of observational skills, thinking and judgement.
5. Science as a means of providing education for leisure.
6. The development of the pupil's self-knowledge.
7. A growing awareness of science in society.
8. A developing understanding of basic scientific ideas.

Once again, of these eight statements, the majority are concerned with the development of cognitive aims (3, 4, 6, 7 and 8). In this case, and somewhat unusually, a number of laudable social aims are also identified (1, 3, 5 and 7). A minority of the aims are concerned with the affective development of the children, (2 and, indirectly, 5). For a project that explicitly acknowledges that the key to its success must lie in the solution of 'the two inter-related problems of ability and motivation', there is a glaring lack of balance between the statements that refer to affective aims and the others. If the successful attainment of long-term cognitive, psychomotor and social aims lies in the initial priming and constant reinforcement of children's motivation, then this philosophy must be unambiguously reflected in the aims and objectives of any derived course. Such *a priori* aims and objectives must permeate all aspects of the curriculum development process: the methods employed to teach the material; the assessment and evaluation procedures; and the content itself. Every stage in the curriculum design (see chapter 3) must be seen to reflect its underlying philosophy, and any that fails to do so must be excised. It is in this sense that the Objectives Model of curriculum development (see p. 47) has been so frequently misapplied. The writing of long-term *a priori* aims is an extremely difficult and underestimated part of the curriculum developer's job, a part to which sufficient attention is rarely paid. So much is this the case, that there is evidence to suggest that some so-called '*a priori*' aims have been retrospectively derived, that is, they have been decided upon after the main content has been written (Nisbet, 1968). Either that, or the curriculum developers have become so conditioned to the inclusion of specific science content into their curricula that they are no longer able to perceive when it not only fails to reinforce their long-term aims, but may be blatantly antagonistic to them.

The early Nuffield Biology (1966) curriculum provides another example of misapplication of the aims and objectives model. One of its stated aims was to 'encourage a respect and feeling for all living things'. It is a little disconcerting to discover in the first textbook, and the one written for 11-year-old children, in a chapter entitled

'How Living Things Begin', instructions to children to break open hens eggs which had been incubated to various stages of development. The children are told:

> after chopping away the surface of the egg, break it into a warm petri dish and push the embryo to the edge of the yolk. The embryo will still be attached to the yolk by blood vessels, so cut these with scissors in order to free the chick completely. The chick dies almost [sic!] immediately and so it will not feel anything now.

Another exercise involves flooding the embryo with warm saline solution to prolong its life so that the heartbeat can be observed. Eventually the embryos are flushed down the sink. How such an exercise, however well it demonstrated the gross foetal stages of vertebrate ontogeny (and it does that very well), could have been equated with the aim of encouraging respect and feeling for all living things is difficult to see. The revised edition of the course, published eight years later in 1974, makes a radical reappraisal of this part of the content. Nevertheless, one is left even now with the impression that the reappraisal was due to the reaction of some of the children rather than to the realisation that this particular part of the course was failing to reinforce one of its major stated aims. We are informed in the second edition of the *Teachers' Guide 1* that 'while some children find looking at the stages in the development of the chick fascinating, others find it distasteful and are distressed by it' (Revised Nuffield Biology, 1974).

Such mismatches between aims and content are relatively easy to adjust, provided that proper evaluation procedures are undertaken. More insidious is the nature of much school and locally produced science curriculum material, which often fails to specify aims and objectives at any stage, either *a priori* or *a posteriori*. It is common to find science syllabuses, particularly those written for mixed ability classes at the junior end of the secondary school which have been developed largely upon tradition and an intuitive feel by experienced teachers for what content is applicable for children of this age and ability.

Just such a syllabus was evaluated by Reid and Tracey (1985). Twelve science teachers in a large coeducational comprehensive school had written a course designed for the children in the first two years of their secondary schooling. They had done so with the best of intentions, but being inexperienced in the theory of curriculum development they had failed to prepare a list of long-term aims to guide decisions on course content. After a retrospective study of the course content, a list of 13 aims was identified and extracted by the researchers, and each member of the science writing team was asked to place these aims in rank order – the aim being perceived as most important was to be placed first on their lists and so on through to the one considered to be of least importance. The mean ranking for all the teachers was calculated. The course was then examined in detail, and the contents allocated to one of the 13 aims. At the completion of the exercise it was possible to rank the aims yet again, this time on the criterion of the importance with which the course treated them. A comparison of the rankings of the two lists of aims revealed a fascinating dislocation. The modest correlation of 0.48 between the time allocated to course aims and the perceived importance of the aims by experienced and competent teachers indicates the extent of the mismatch. It appears that the intuitive nature of the exercise was an inadequate substitute for a more disciplined approach. It is particularly interesting to note that the retrospectively derived aim that appeared to be most important to the teachers was 'to develop in their pupils an interest and enjoyment in science'. This aim is very close to what in general terms we think of as 'motivating' pupils. It probably comes as no surprise to find that this aim received only half the time allocation afforded to some of the cognitive aims, in particular 'knowledge of facts and concepts concerning the environment', and, although it is acknowledged that the time spent on any particular aim is not a cast-iron reflection of the influence the teaching is likely to have, it must be considered one of the major criteria influencing children's perception. When the attitudes of the children in their first three years in the school were measured, it was found that they were less favourable at the end of the course than they had been at the beginning, and this was especially marked with the girls. Moreover, it was a deterioration in the enjoyment of science which was most marked. The intuitively derived science content had failed to reverse a commonly recorded phenomenon amongst secondary school children (Whitfield, 1979).

It is, of course, impossible to say what the outcome would have been had the course genuinely reflected the desire of those teachers to fulfil their main ambition of promoting active interest and enjoyment in science. There is no escaping the fact either that deterioration in children's attitudes both to school generally and to subjects in particular is a well-documented phenomenon. Hadden and Johnstone (1983) demonstrated an erosion of favourable attitudes to geography, arithmetic and mathematics in Scottish schoolchildren, although the erosion was more pronounced in science. It is unlikely that subjects other than science have a more enlightened attitude to the priming of motivation in children. In this respect science has two major advantages. It lends itself to the application of the aims and objectives model of curriculum development, which makes evaluating the success of a science course a viable procedure, and some of it at least is inherently interesting. This should provide optimism that better courses for the less able and the underachieving are possible.

THE UNDERACHIEVER'S COGNITIVE RESPONSE TO SCIENCE

The accent in this chapter is on motivating the underachieving child, but reference to figure 2.2 makes it clear that motivation cannot be considered in isolation from the cognitive state of the child. Every child, regardless of ability, comes to school knowing some science. Every child has been hurt at some stage before the age of 11. Children know what 'heat' is because they have been burned by climbing into too hot a bath at some time, and they know what 'balance' is because they have fallen off their bicycles. They hear their teachers using words with which they are quite familiar. But less able children are not so sensitive as their more able peers to the subtle way in which the same words may be used in different contexts. Less able children are not alerted sufficiently to the distinction between the *everday* reality of what 'heat' is, and the *scientific* reality of what 'heat' is. This is by no means as unusual a condition as might be supposed. Science jargon is replete with metaphor. How many graduate biologists, chemists or physicists perceive ionic 'pumps', chemical 'pathways' or electrical 'fields' as 'scientific' explanations of natural phenomena, for example? One difference between scientifically literate adults, and less able, scientifically naive children is, of course, the extent of the gap separating their personal reality of science and the scientific reality of that science. Science and technology teachers are separated from their less able pupils by a chasm of experience, so they live in different perceptual worlds from each other.

Johnson Abercrombie (1960) has pointed out how prior knowledge can affect our interpretation of phenomena. She quotes Breuil on the Quaternary cave paintings, such as 'the bellowing bison' (figure 2.3), as supposing the artists 'were hunters who became so familiar with the behaviour of the animals in the field that they were able to take back to the caves snapshots in their mind's eye of the beasts in characteristic poses, which they rendered in paint with extraordinary skill'. An alternative explanation is suggested by Leason, an artist who had made drawings of dead animals. Leason says of the 'bellowing bison' of Altamira, 'the spectacle of some farmyard cow rising lightly on all four hoof tips and emitting a bellow, is too much for the imagination'. As far as he is concerned, the cave artists were simple copying the outline of the dead animal

(a) A 'bellowing bison' after Abbé Breuil. Taken from a tracing of one of the Altamira cave paintings.

(b) A dead sambur hind after P. Leason. Drawn from a photograph of a dead animal.

Figure 2.3 Two perspectives of cave drawings (adapted from M. L. Johnson Abercrombie, 1960; reproduced with permission)

reposing on the cave floor after it had been ignominiously dragged there by the hunters. As Abercrombie points out, we shall probably never know which hypothesis is correct. But 'according to the preconception that the artists' models were alive or dead, observers notice or neglect certain parts of the picture and may interpret others differently'. Head (1985) shows how experience can affect pupils' perception of scientific phenomena, and both he and Driver (1983) stress the tenacity with which children can hold on to their intuitive notions about scientific phenomena. From a series of systematic studies, Driver lists some of the 'alternative frameworks' children hold about science. Over 10 per cent of 11 year olds in Britain believe, for example, that sun, wind and fire are 'living things'; it is not difficult to imagine why this might be once we have become alerted to the possibility. Gas showrooms advertise 'living fires', and children's literature anthropomorphises 'friendly' suns and 'cruel', 'biting' winds.

The yawning chasm between what teachers bring to their teaching and what underachieving children bring to their learning in terms of intuitively held beliefs about science is, then, exacerbated by a host of factors. It is the function of this book to begin the process of teasing these out prior to the selection of specific programmes of school work. An obvious starting point, says Ausubel (1968), is to discover what the pupil already knows. 'Ascertain this and teach him accordingly'. Driver (1983) puts it this way:

After all, if a visitor phones you up explaining he has got lost on the way to your home, your first reaction would probably be to ask 'Where are you now?' You cannot start to give sensible directions without knowing where your visitor is coming from. Similarly, in teaching science it is important in designing teaching programmes to take into account both children's own ideas and those of the scientific community.

PART II
Designing a Curriculum for the Underachiever

Introduction

It is now more than 35 years since Ralph Tyler (1949) identified four key questions that teachers and curriculum designers should ask when embarking on a curriculum planning exercise:

1. What educational purposes should schools seek to attain?
2. How can learning experiences be selected to attain them?
3. How can these experiences be organised for efficient instruction?
4. How can the effectiveness of the process be evaluated?

Since that time, Tyler's proposals have been elaborated by curriculum theorists such as Taba (1962) and Wheeler (1967) into what has become known as the Objectives Model of curriculum

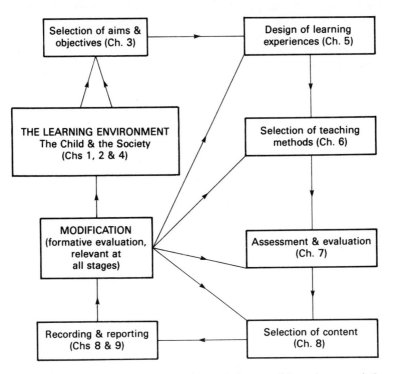

Figure 3a The components of a simple curriculum model, and some of the interactions between them

planning. Whilst extreme versions of this model (Mager, 1967; Popham and Baker, 1970a,b) have been criticised on the grounds that they 'dehumanise' learners and trivialise learning (MacDonald Ross, 1973; James, 1968; Stenhouse, 1975), the basic notion that the curriculum can and should be planned systematically is now well established. The figure on p. 41, representing the processes of systematic curriculum planning in simple diagrammatic form, makes clear the sequential and cyclic nature of the process: aims and objectives determine teaching/learning methods and content, and these provide the basis for the design of assessment and evaluation procedures, which, in turn, furnish information on which to base decisions concerning curriculum modification and renewal. This diagram provides a structure for part II of this book, with a chapter focussing on each of the major steps in the cycle. At this stage it is important to point out that we are not advocating strict adherence to the principles of planning by behavioural objectives. We do not want what Dewey (1916) called 'the tyranny of ends'. What we are advocating is clarity of thought regarding the purpose and orientation of the science curriculum, such that there is harmony between that purpose, the curriculum experiences of the children, and the assessment and evaluation procedures employed by the teacher to provide feedback on the success, or otherwise, of the curriculum enterprise.

3

Aims and objectives

AIMS

As indicated in the Introduction to part II, any set of proposals for the science education of young people should, logically, begin with a statement of aims. Such aims derive from the educational philosophy underpinning the curriculum, in particular the issues discussed in chapter 1. It was stated there that a course for scientific literacy should not exclusively adopt any one orientation. Rather, it should derive its design principles from, collectively, the structure of the disciplines, consideration of the learner and the nature and needs of society.

Some pointers towards the kind of science curriculum we ought to provide can be extracted from a consideration of the kind of society our children will inhabit when adult. We need to ask questions such as 'What kind of society are we preparing them for?' and 'What major societal changes are likely to occur during their lifetimes?' There is no doubt that, for the first time in history, we are preparing children for adult life in a society of which we can, at this stage, know very little. Because of the rapid pace of change, society will present citizens with constantly changing intellectual demands. Thus, we need to develop attitudes favourable to lifelong education and to retraining needs. It may well be that those about to enter employment during the next decade will need to retrain several times during their working lives. Thus, attitudes enabling them to cope with these demands may well be far more important than any specific content. Because societal changes are often science and technology based, future citizens will need to understand science and technology, and the way in which they interact with society, if they are to act as intelligent and responsible participants in the decision-making processes of a democratic society. It is important that all citizens can form views about the desirability or undesirability of particular technological changes, because those who are scientifically and technologically illiterate are at the mercy of unscrupulous propagandists.

Of course, the extent to which we can achieve such goals will vary enormously across the ability range and will be subject to a whole range of additional factors beyond our control. Factors such as gender, race and social class will all play a part. However, we believe very strongly that we can take all children some way 'along this road', and that this should be a major goal of the science curriculum. As far as less able and underachieving children are concerned, our proposed goal is heightened personal awareness. To us, this has two major aspects: heightened awareness of the natural world and heightened awareness of the created environment, which, in Western society, is science and technology dependent. The achievement of this heightened personal awareness is closely associated with increased personal satisfaction – taking pleasure in understanding and appreciating the natural and created environment. In chapter 2 we discussed the interrelationship of the cognitive and the affective and concluded that, for our target group of children, the affective domain was the more significant and should have curriculum priority. Consequently, we would argue that the cognitive goal of heightened personal awareness, necessary to the ultimate goal of achieving a fully literate citizenry, is best approached from the affective goal of increased personal satisfaction. Thus, we see the major goals of science education as falling under three main headings. In order of curriculum priority, these are:

1. learner-centred aims (concerning issues such as motivation and the development of attitudes and feelings, and the enhancement of self-image),
2. society-centred aims (concerning the interaction of science and society), and
3. science-centred aims (related to the structure and methods of science).

A curriculum oriented towards the attainment of universal scientific literacy would have aims in all three of these areas. In keeping with our declared intention of setting out guidelines and principles of procedure, we advance the following as a tentative set of aims for such a curriculum:

Learner-centred aims
- the development of skills of communication of ideas and feelings
- the establishment of a sense of personal and social identity and adequacy, including confidence in one's ability to tackle and solve problems
- the attainment of self-esteem through successful experience and learning
- the formation of mature relationships with peers and adults; this includes the willingness to accept responsibility for one's actions and the ability to work co-operatively with others
- the encouragement of science as a worthwhile leisure pursuit, possibly as science-based hobbies and interests.

Society-centred aims

- a basic understanding of the nature of advanced technological societies and of the complex interactions between science, technology and society, taking into account contemporary world and national issues, together with local issues and historical perspectives
- an appreciation that, in decision making, scientific and technological criteria have to be balanced against economic, ethical and social considerations.

Science-centred aims

- the attainment of knowledge and understanding of a range of scientific concepts, facts and theories through systematic study and experience of phenomena
- the acquisition of a range of cognitive and psychomotor skills associated with the practice of science
- the ability to utilise scientific knowledge and scientific processes for the understanding and exploration of physical phenomena and for the solving of problems
- the attainment of a scientific way of looking at the world, together with some understanding of how the scientific approach resembles and differs from the approaches of other disciplines.

There is no doubt that traditional science curricula, and even the more progressive Nuffield-inspired courses, have placed primary emphasis on category 3 aims. There are signs of some shift towards category 2 aims, especially in recent publications of the ASE and SSCR, but – as yet – there is little evidence of a shift towards category 1 aims. We regard this shift in priority as desirable for all children, but crucial to curriculum success for the underachiever. Only by refocussing our attention to take account of children's personal and affective needs can we begin to raise levels of attainment in the cognitive domain.

It is an interesting contemporary fact that some children who underachieve in school attain very high standards of achievement when playing computer games. Perhaps the key to that attainment, as we have argued the key to all attainment, is the maintenance of high levels of motivation. We should ask, then, whether teachers have anything to learn from computer gaming methods. Certainly they make frequent and significant use of reinforcement – 'learners' are rewarded in a variety of ways for achieving higher levels of performance. We need to adopt similar approaches and more frequently provide for both immediate knowledge of results and praise and support for learners; specific examples of how this is currently being accomplished by computer assisted learning (CAL)

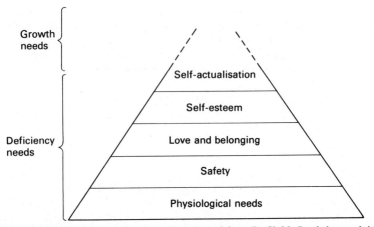

Figure 3.1 Maslow's hierarchy of needs (adapted from D. Child, *Psychology and the Teacher*. London: Holt, Rinehart & Winston, 1973)

are given in chapters 5 and 6. A much more subtle, and probably more influential, set of motivating factors are the learner's aims and intentions, needs, wants and aspirations. This is, of course, a very complex set of issues and is rooted in personality development, but some clues are provided by a consideration of Maslow's (1970) hierarchical needs pyramid (see figure 3.1). Basic human needs are arranged as a hierarchy. Needs at a lower level must be satisfied before we can attend to those at a higher level, and progress at higher levels is enhanced when lower level needs have been thoroughly met. It is central to Maslow's theory that the affective needs lower in the hierarchy must have been thoroughly met before any significant progress in the cognitive domain can be made. The lower levels of the hierarchy reinforce our earlier assertions that the creation of a stable emotional environment is essential to cognitive growth for underachievers. At a higher level, Maslow identifies two sets of needs associated with self-esteem: the desire for competence (adequacy, independence, freedom), and the desire for recognition (reputation, prestige, importance, appreciation). Again, these ideas reinforce our earlier assertions that such needs must have curriculum priority for 'our children'. If we are to reach higher levels of the cognitive hierarchy then we must encourage the development of each child's self-esteem and self-actualisation.

Weiner's (1979) attribution theory and the more global theory of locus of control (Rotter, 1966) have been used in a variety of settings to investigate children's own explanations of success and failure in academic settings in terms of internal (ability, effort) and external (luck) factors. What is apparent from these studies is that academically successful children more frequently feel that they control events, whereas less successful learners commonly admit to feeling that external factors control events concerning learning. If we turn this principle around, we have grounds to believe that a feeling of 'being in control' is a necessary prerequisite to academic success. If learners feel that success in learning is in their hands and is not just a matter of luck, they are more likely to learn successfully. Children who believe that success derives from personal effort are more likely to persist in their efforts, provided that they see the goal as worth while. It is our belief that giving children a measure of responsibility for devising their own learning strategies will increase their sense of control of events and will lead to higher levels of attainment. Giving children some choice about the topics they will study, and the order in which they will study them, will have a similarly positive effect. In summary, we need to alter the locus of control from the teacher to the pupil. Or, more importantly, we need to alter the children's perceptions of where that control lies. We are certainly not arguing that realigning the perceived locus of control is a guarantee of academic success, but we are saying that it is a sensible place to start. We believe that we can go some way towards the creation of more favourable learning conditions by emphasising category 1 aims, and by allowing a measure of personal control of the curriculum, through the adoption of negotiated content and by using learner-controlled learning methods, especially CAL and project work.

OBJECTIVES

Whilst curriculum aims are useful, in that they provide an orientation for the curriculum, the Objectives Model – or 'rational curriculum planning', as its hard-line devotees call it – requires that they be operationalised, translated into the form of objectives, which can provide the teacher with precise guidance on how to proceed in the classroom. These objectives must be stated not in terms of what the teacher intends to do, but in terms of what the learner will be able to do after the teaching/learning process is completed. The purpose of education is not to have the teacher perform certain activities, but to bring about certain significant and desirable changes in the learner's patterns of behaviour. Objectives framed in terms of curriculum content only are also unacceptable, for they do not indicate what the learner is required to 'do' with that content – repeat it, understand it, analyse it, criticise it, or what? Thus, a syllabus is not a set of learning objectives. A syllabus does not indicate what a learner will be able to do, or say, or perform, after the learning experience is completed and it is, therefore, insufficient guide to planning learning experiences and to evaluating them. On this latter point, a number of curriculum writers go so far as to assert that accurate assessment of the information, concepts and skills which children have acquired cannot be made unless the objectives are expressed in terms of an intended change to be brought about in the learner. That intended change should be expressed in terms of measurable learner behaviour. 'A satisfactory instructional objective must describe an observable behaviour of the learner' (Popham, 1970).

> The most important characteristic of a useful objective is that it identifies the kind of performance which will be accepted as evidence that the learner has achieved the objective ... it specifies what the learner must be able to DO or PERFORM. (Mager, 1967)

To assist the processes of design and assessment, Wheeler (1967) suggests that general aims, such as those specified above, should be analysed to provide, successively, ultimate goals, mediate goals, proximate goals and specific objectives. Ultimate goals are general aims expressed as 'patterns of behaviour'; mediate goals are behavioural goals 'along the way' to ultimate goals and are necessary because learning of one kind of behaviour is an essential prerequisite to the learning of a higher-order behaviour; proximate goals are the targets of units of work; specific objectives are the learning outcomes of particular encounters with particular children and are 'best expressed' in the behavioural terms familiar from Mager's (1967) work. That is, they should comprise three parts:

- terms describing the behavioural situation
- measurable performance terms
- qualifying terms that describe the level of sophistication needed for acceptable performance.

Whilst welcoming the hierarchical, modular design implicit in Wheeler's proposals, we do not accept that it is necessary, or even desirable, that teachers should specify curriculum goals in this kind of detail. It is our view that the degree of specificity necessary in the formulation of curriculum goals is that which is necessary to guide an experienced teacher in planning, executing and evaluating the curriculum. Thus, curriculum aims should provide such a teacher with the 'flavour' of the course – its overall 'end-in-view' as Dewey (1916) calls it – and curriculum objectives should indicate the kinds of experiences the children will engage in and the kinds of learning outcomes to be expected. They need not specify precise, measurable behaviours of the Magerian type. Three important criticisms of the Objectives Model of curriculum design are embedded in this stance.

First, that in concentrating attention on the specification of learning outcomes, it fails to recognise that many worthwhile curriculum activities have no specifiable end result. Thus, it confuses education, which is always unfinished, on-going, unpredictable and uncertain, with training, which can be completed and is readily seen to be so by observational methods. Stenhouse (1975) clarifies this distinction and delineates education into four processes: training, the acquisition of skills; instruction, the learning of information; initiation, familiarisation with social norms and values; and induction, introduction into the thought systems of the culture. For instruction and most forms of training the objectives approach is adequate. For initiation and induction, however, it is totally inadequate. Indeed, Stenhouse goes so far as to say that 'education as induction is successful to the extent that it makes the behavioural outcomes of students unpredictable'. What is needed in these cases is what Eisner (1969) calls 'expressive objectives'.

> An expressive objective describes an educational encounter. It identifies a situation in which children are to work, a problem with which they are to cope, a task in which they are to engage; but it does not specify what from that encounter, situation, problem, or task they are to learn.... The expressive objective is intended to serve as a theme around which skills and understandings learned earlier can be brought to bear, but through which those skills and understandings can be expanded, elaborated, and made idiosyncratic. With an expressive objective what is desired is not homogeneity of response among students but diversity.

Second, the Objectives Model seriously underestimates the complex relationship between means and ends, and draws a distinction between them that is not always valid. Whilst certain goals can be achieved by the adoption of particular actions, many worthwhile goals are part of the action itself. For example, developing a child's capacity to employ evidence in the formulation of hypotheses is a worthwhile end, but it is also the means to another end related to the particular subject matter. Moreover, to regard it as an end for which suitable means have to be found is absurd – one can only learn to formulate hypotheses by formulating hypotheses! In other words, the end and the means of achieving that end are identical. The Objectives Model fails to recognise that multiple outcomes derive from certain single experiences and, conversely, many experiences may be needed to attain certain single outcomes. For convenience, learning goals can be crudely separated into two categories: those that are the specific goals of a particular learning episode (or series of such episodes) and are related to particular content, and those that are general goals for all lessons, regardless of content. The model is able to accommodate the former but not the latter category of objectives.

Third, the Objectives Model fails to draw a distinction between the learning goal and the evidence that we seek, and will accept, as an indication that the goal has been attained. Whilst the latter – perhaps better termed the performance objective – can be expressed in terms of expected learner behaviour, the former cannot. What is needed is a set of objectives sufficiently specific to guide the teacher in planning the learning experiences and a set of precise, behaviourally expressed assessment or performance objectives to assist the design of an effective and efficient scheme for the assessment of learning and the evaluation of curriculum experiences (this latter procedure forms the basis of chapter 7). The distinction we are attempting to draw between learning goals and evidence of their attainment is similar to that drawn some years ago by White (1971), who insisted that we should distinguish between (i) 'objectives which themselves consist in pupils behaving in certain ways', and (ii) 'objectives whose attainment is tested by observing pupils behaving in certain ways'. It is our contention that White's category (i) objectives are appropriate to instruction and training, whilst his category (ii) objectives are more appropriate to induction and initiation.

In summary, what we are arguing for is that some science education goals should specify output (intended learning outcomes) whilst others should specify inputs and processes (learning experiences). Stenhouse (1975) illustrates this crucial distinction by describing how one learns to play chess:

> Certain simple skills are necessary to begin: we must know the basic moves. These can be dealt with through objectives. Standard openings are more difficult. First a learner needs to see a use for them, and then to understand the principles of each. He needs to explore them as they arouse his interest. For much of the time, beyond this, we cannot tell the learner exactly what to do. We can advise him on principles, we can help him analyse his successes and failures and the games of others. But he must move autonomously, he must act under his own direction if he is to learn. And note, sometimes he will beat the teacher, but he will never arrive at the point where he has learnt all.

The remaining part of this chapter is concerned with the itemisation of objectives, both instructional and expressive, deriving from two major sources:

- a consideration of the needs and interests of the learners, particularly in relation to the need to motivate to enhance self-image, and

- a consideration of the nature of science and technology, their interaction with society and their mode of operation.

For convenience we have divided these objectives into three major categories: those concerned with the acquisition of scientific knowledge, those concerning the understanding and utilisation of scientific processes, and those that focus on the promotion of certain attitudes. It should be apparent from the preceding arguments that these categories have been listed in order of reverse priority. This is not mere perversity on our part, it just happens to suit our purposes better at this stage!

Knowledge objectives

Knowledge objectives derive from two principal sources: from the proposed content of the curriculum (certain concepts, facts and theories) and from considerations in the philosophy of science. Content is discussed in some detail in chapter 6, and nothing is to be gained by detailed discussion of these objectives here. There are, however, a number of philosophical issues that warrant attention at this stage.

In all curricula, implicit messages about the nature of science are evident in the choice of specific content and learning methods. It is our contention that such messages should be made more explicit and used as a focus in planning curriculum experiences. If children are to acquire a proper understanding and appreciation of science and scientific activity, it is necessary that philosophical considerations are afforded a more prominent role in curriculum design. Contemplation of the vast and complex literature generated by philosophers reveals no single, universally accepted view of science, though there is a measure of agreement relating to the theory dependence of observations and the relationship between observation and theory (Hodson, 1986c).

At least two objectives derive from the analysis of the role and status of observation in science:

- recognition that observation is unreliable and theory dependent
- realisation that the techniques of scientific observation have to be learned.

It also follows that we need to take account of children's existing conceptual frameworks, since they profoundly influence the observational work they can undertake and their success in acquiring new ideas. Thus, it is necessary to provide 'training' in observation techniques and a rich and adaptive learning environment, matters that are discussed in detail in chapters 4 and 5.

More problematic than the role and status of observation is the role and status of theoretical knowledge. In order to bring about a proper understanding of science it is necessary that the role of theory is made apparent to the learners. Whilst it is clear that its role is to explain phenomena, it is not clear whether that explanation has the status of a true description of the world (a realist position) or of a convenient fiction (an instrumentalist position). Hodson (1985a) has argued that these extreme positions can be avoided by adopting a developmental model, in which every major theory in school science passes through several stages:

- tentative introduction as a model (with frequent use of pupil ideas)
- a search for supporting evidence and for falsifying evidence through observation and experiment, leading to choice of the 'best' model
- further elaboration of the chosen model into a theory, through refinement of concepts and establishment of quantitative relationships
- acceptance of the theory into the body of scientific knowledge
- use of the sophisticated theory to explain phenomena
- testing of the theory's predictions and application of the theory in new situations.

At each stage we need to ensure that children appreciate the role and status of the theory (or model) under consideration. Such considerations lead directly to concern with the methods of science – the ways in which scientists investigate phenomena, generate new knowledge and test it for adequacy against experimentally gathered data. The inability of philosophers to describe a single, universally accepted method of science does not imply that science has no methods. Scientists employ an approach that depends crucially on the matter under investigation, the existing theoretical knowledge and the range of experimental techniques available in the field. In doing so they use the most appropriate selection of processes from a range of those available. It should be noted that the 'processes of science' refer not to the manipulative skills employed in carrying out laboratory exercises, such as using a microscope or a top pan balance, but to the strategic skills of conducting a scientific investigation.

Process objectives

Recent publications from the SSCR (1984) and the DES (1985a) emphasise the importance of teaching the processes of science. We want to go further than this by emphasising the priority of processes. To that extent we are in sympathy with *Science 5–13* (Schools Council, 1973b), which could, in a sense, be regarded as a content-less course (Harlen, 1978). In an attempt to identify the processes of science, the American Association for the Advancement of Science asked scientists to say what is involved in day-to-day scientific research. The responses were simplified and categorised into 13 processes:

Observing	Using space–time relationships
Classifying	Interpreting data
Communicating	Formulating hypotheses
Predicting	Controlling variables
Inferring	Defining operationally
Measuring	Experimenting
Using numbers	

These 13 processes have been used as the basis for science curriculum development, in particular the development of the American elementary school project *Science – A Process Approach*. In our view the list is insufficiently detailed to guide teachers in the construction of their own science curriculum and should be replaced by the more extensive list of 21 processes (table 3.1) provided by Hodson and Brewster (1985). This list is sufficient to guide the design of learning experiences and the construction of an effective assessment and evaluation scheme (see chapter 7). We see the use of these process objectives, together with the application of relevant knowledge from the content objectives, as providing the framework for organising a systematic curriculum planning procedure.

Attitude objectives

The third category of objectives necessary for systematic planning, and those to which we afford curriculum priority, are those concerned with attitudes and feelings. If we think specifically in terms of science, these attitudes would seem to divide into two major categories:

- attitudes to science
- scientific attitudes.

The first of these categories is the traditional one of attempting to foster positive attitudes towards science as a basis for occupation and leisure pursuits, and as a social and economic force. The second category represents 'the motivation which converts [scientific] knowledge and skill into action and refers to a willingness to use

Table 3.1 *The processes of science*

---

1. *Planning investigations*
   - P1   Identification and clarification of problems (asking appropriate questions)
   - P2   Formulation of hypotheses
   - P3   Selection of suitable tests of hypotheses
   - P4   Design of experiments:
       - (i) analysis into component steps
       - (ii) identification and control of variables
       - (iii) selection of appropriate procedures and apparatus
       - (iv) identifying safety issues

2. *Performing investigations*
   - P5   Accurate observations of objects and phenomena
   - P6   Selection of appropriate measuring instruments
   - P7   Accurate measurement
   - P8   Describing and reporting observations in appropriate language: (i) qualitative, (ii) quantitative
   - P9   Safe use of laboratory equipment
   - *P10 Performance of routine laboratory operations
   - *P11 Performance of specific techniques
   - P12   Carrying out familiar and unfamiliar procedures in accordance with written or verbal instructions
   - P13   Methodical and efficient working

3. *Interpreting and learning from investigations*
   - P14   Processing, manipulating and organising experimental data
   - P15   Presentation of data in a suitable form
   - P16   Analysis and interpretation of data (recognising trends, sequences and patterns)
   - P17   Extrapolation of data and generalisation
   - P18   Making sense of data by reference to relevant theory
   - P19   Drawing conclusions (including the relationship between hypotheses and interpreted data)
   - P20   Suggesting modifications and improvements for further work

4. *Communication*
   - P21   Preparing and communicating an oral or written account or report in a suitable form, taking into account both content and audience

---

*Specific items in P10 and P11 would have to be listed

scientific procedures and methods' (Gauld, 1982). In other words, they are the attitudes associated with the successful study and practice of science. They are also the necessary learner attitudes underpinning the principal psychological stance taken in this book – that the major cognitive task facing the teacher is to bring about a shift from personally held beliefs to the accepted scientific paradigm. If this is the case, then teachers will need to foster in pupils a set of attitudes – such as open-mindedness, self-criticism, curiosity, etc. – that will facilitate such a shift. What is interesting is that many of these attitudes and characteristics complement, or even rein-

Table 3.2 *Attitudes and interests reinforced by science*

A1 Independence of thought and self-confidence
A2 Capacity for self-motivation and acceptance of responsibility for one's own learning
A3 Perseverance and tenacity in the face of difficulties
A4 Intellectual curiosity
A5 Tolerance of the views of others
A6 Self-criticism; also a willingness to criticise and be criticised by others
A7 Co-operation with others, consisting of
  (i) carrying out tasks together and
  (ii) a willingness to pool data and ideas
A8 Open-mindedness: willingness to change one's mind in the light of new evidence; willingness to suspend judgement if there is insufficient evidence
A9 Appreciation that most issues and problems can be approached from a variety of perspectives
A10 Honesty and integrity in carrying out and reporting experimental work
A11 Willingness to predict, speculate and take 'intellectual risks'
A12 Acceptance of scientific inquiry as a legitimate way of thinking about issues and problems
A13 Enthusiasm for science
A14 Informed and healthy scepticism based on recognition of the limitations of science; this would include the capacity to resist claims unsupported by evidence or theory
A15 Recognition of the role of science and technology in shaping society and our material well-being
A16 Application of science problem-solving skills to everyday situations
A17 Adoption of science-related interests, e.g. keeping animals, computing, 'radio ham', reading science fiction

force, the scientific attitudes identified by Gauld (1982) and Schibeci (1984). We see the list of attitudes in table 3.2 as comprising the most essential checklist for the selection of specific lesson content and the design of appropriate learning experiences. It is with these objectives that we see the foundation being laid for a more relevant and a more successful curriculum for those who currently achieve so little in school.

We wish to emphasise here that whilst we are approaching these objectives through the science curriculum, we believe that they are of value outside the laboratory, and that they are transferable. Thus, we see the science curriculum, generally, as a vehicle for the general personal development of children (what we called increased personal satisfaction) and, specifically, for the heightening of personal awareness of the natural and the created environments. In these respects, we see that our curriculum proposals are significantly different from those of most (all?) other science curricula. The learning environment through which such objectives are best pursued and through which children achieve heightened personal awareness and enhanced self-image – in other words, the necessary conditions for the attainment of scientific literacy – is discussed in the following two chapters. Only when sound pedagogic principles have been established can we begin to choose learning methods and content appropriate to our eventual goals.

It is our belief (faith, even) in the power of an approach based on a reordering of priorities and on careful organisation that leads us to oppose the determinism of many teachers, a determinism that condemns large numbers of children as failures. That determinism is rooted in an acceptance of low ability, poor reading, inadequate social background, race, gender, and all the other factors that influence educational attainment as explanations of academic failure. As soon as these factors are accepted as explanations, the determination to ensure success evaporates and failure is inevitable, as teachers see them as beyond their control. If they are identified merely as factors and, in particular, as factors for which we can compensate, the optimistic picture that all will learn successfully can be retained. Curriculum planning then becomes a matter of organising the learning environment and the learning experiences to attain the common curriculum goals in accordance with the particular characteristics of the individual learners.

———4———
The internal learning environment

The aims and objectives model of curriculum development has so far enabled us to ask vital questions about the philosophy underpinning the concept of 'science for all', especially for less able and underachieving children, which in turn provides invaluable guidelines for syllabus content. Knowledge of the psychology of less able and underachieving children will also affect the selection of specific content and method within the confines of this philosophical stance. But in addition, it is becoming more evident that the interface itself between content, method and child is of great importance. Content and child interact through the teacher in the laboratory by the process known as communication. In this chapter we look in more detail at the psychology of underachieving children, and in the next at the process of communication itself.

REASONS FOR UNDERACHIEVEMENT

The underlying reasons for underachievement have always been a cause for concern and dispute amongst educationalists. At the turn of the nineteenth century it was held that underachievement was a direct result of low innate ability. In the 1890s Binet attempted to produce tests which would be able to select out those children unable to cope with ordinary classroom work in the schools in Paris, and the process has been perfected until today there are literally hundreds of reliable and valid tests designed to measure different kinds of mental agility (NFER, 1986). By the middle of the twentieth century however, the idea of intelligence as an immutably inherited genetic trait was being vigorously challenged by writers such as Douglas (1964) and Floud, Halsey and Martin (1957). Benn and Simon (1970) stress the influence of cultural background on achievement in schools by quoting the fact that the chances of a middle-class boy living in Cardiganshire getting into a grammar school in the 1950s was 160 times greater than those of a working-class girl from West Ham. In the late 1960s and 1970s, Bernstein (1973) moved the debate into a new arena when he presented his socio-linguistic theory of underachievement. In this he claimed that underachievement by the 'working classes' was intimately connected with the way in which working- and middle-class speakers misinterpreted each other's intentions, particularly in institutions that were themselves middle-class oriented such as schools. Whilst elements of his theory have been criticised (e.g. Edwards, 1976), there can be no doubting the influence that Bernstein has had upon the value educationalists, and science educators in particular, now place on the importance of communication in the learning process. HMI reports of secondary school inspections are often critical of this aspect of the learning environment:

> the amount of talk generated as part of the lesson is often limited. Most classroom conversations consist of questions from the teacher evoking short answers from the pupils, and few schools consistently encourage pupils to develop arguments; to formulate as well as to answer questions; and to articulate their ideas through more open discussion. Given the opportunity, pupils demonstrate that they are capable of engaging in these activities and that they enjoy doing so. (DES, 1984)

HMI stress the responsibility that individual teachers must take for improving not only oral but also written communication by children:

> The ways in which teachers use and stimulate their pupils' use of language in the classroom are among the most frequent aspects of teaching skill seen as critical to the success of individual lessons... H.M.I. reports frequently refer to teachers using language in ways which limit pupils' responses... it is the teacher in the classroom who must use judgement, sensitivity and skill to ensure that children's powers of oral and written expression, as well as their ability to listen, are enhanced and encouraged. (DES, 1985b).

The process of communication is much more than the simple transmission of information from one person to another. Scheflen and Scheflen (1975) define it as 'all behaviours by which a group forms, sustains, mediates, corrects and integrates its relationships'. The word derives from the Latin *cum* and *unus*, meaning literally 'with oneness', and this is reflected in such phrases as 'the community spirit'. In the school laboratory, communication is the process by which interaction between teacher and pupil, and between pupil and pupil, is facilitated. It carries with it implications of mutual psychological satisfaction; it is then somewhat of an intangible concept. It is possible to see the overall process of communication as being made up of a number of more or less

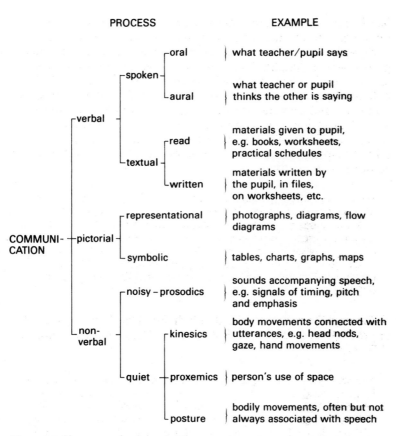

Figure 4.1 The range of communication processes

discrete parts (figure 4.1), but it must not be forgotten that the whole is greater than the sum of these individual parts. In the next chapter we shall extract and focus upon three areas in the communication process that have shown themselves to be specially susceptible to research, and specially relevant to the teaching of less able and underachieving children. These are: spoken language itself (but especially that of the child), children's reading materials, and children's writing. Meanwhile, in this chapter, we shall concentrate on individual differences in the mental development of children, and their sensitivity to non-verbal communication.

The constructivist school of psychology, which argues that all knowledge is constructed by individuals as they interact with their environment in an effort to make sense of it, consistently argues the role of communication in coming to understand the nature of the child's current understanding of science (Osborne & Wittrock, 1985). The first step in 'reconstructing' (hence 'constructivist' psychology) any scientific misconceptions that might be held by the child is to understand the precise nature of those misconceptions, and the context in which they are held. This can only be done by carefully observing what the child does, says and writes. Hence opportunities have to be made that will facilitate the communication of this information. Once revealed, the argument goes, appropriate, individualised learning environments can be devised which will teach the child more efficiently than the non-tailored methods currently in use. There is one flaw in this argument. This is the presupposition that the child can ultimately – provided appropriate teaching methods are adopted – come to an appropriate understanding. In other words, it takes no account of the possibility that there are certain scientific concepts that children, at certain stages in their lives, are incapable of understanding, no matter what teaching strategies are applied. The best endeavours to communicate will founder unless what is being communicated is communicable. A prime tenet of the learning process is that 'nothing suceeds like success', and the coverse is just as much a truism – that nothing is more likely to lead to failure than constant failure itself. It is to this problem of the learning environment, the identification of the intellectual ability and stage of development of the learner, that attention must first be directed.

INDIVIDUAL DIFFERENCES IN INTELLECTUAL DEVELOPMENT

Beginning in the 1920s, the Swiss psychologist Jean Piaget observed and interviewed children, presenting them with intellectual tasks and recording their answers. In general, Piaget found that intellectual development took place in four main stages. In this book we are concerned with children of 11 years of age upwards, and discussion is therefore confined to the last three of these stages (see figure 4.2). The trend is clear, and at this level of generality, quite unequivocal. As children get older they become more capable of dealing with intellectual abstraction and purely symbolic relationships between phenomena.

A recent study by Beveridge (1985) provides an apposite example. He showed that the age of his subjects was a vital determining factor in the way they were able to account for the evaporation of water from a heated pan. Working with 270 5-, 7- and 9-year-old children, he found that increasing age was associated with a tendency to account for the process less intuitively. In other words, the older the

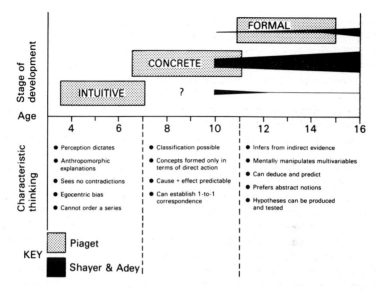

Figure 4.2 Stages of cognitive development (after Piaget, 1950, and Shayer & Adey, 1981) associated with the main thinking characteristics by age

children were, the less likely they were to explain the disappearance of the water in terms of a 'gut' feeling that somehow or other it must be related to the nearest object, namely the pan in which the water was being heated! Explanations shifted from the water is 'going into the pan' to its 'turning into steam'. The younger children, even when they were given lessons related to the inadequacy of the 'absorption' hypothesis or the adequacy of the 'steam' hypothesis, could not escape from their first intuitive thoughts. Indeed, the 7 year olds who were given lessons on the inadequacy of the 'absorption' hypothesis actually used this hypothesis more frequently as an explanation of the phenomenon than the control group. It is as well to remember that, at this stage of cognitive development, temporal contiguity is used by children to impute causality in novel situations (Mandelson and Shultz, 1976).

Science usually demands some intellectual objectivity on the part of the learner, which is one of the reasons why it is considered such a difficult subject. Reference to figure 4.2 indicates that Piaget envisaged some overlap between individual children and the ages at which the different stages of development are successfully achieved. Nevertheless, classical Piagetian psychology would have us believe that, by the time children have reached the age of 11, not only have they passed completely through the intuitive stage of thinking, but they are well on the way to being able, and indeed often preferring, to think in the abstract. In this last of the stages of cognitive development, the child is capable of formal reasoning, which does not depend on the concrete crutches of everyday experience.

Shayer and Adey (1981), in a survey in England and Wales of some 12,000 pupils from a variety of secondary schools, have caused this idea to be re-examined. Using specially developed 'Science Reasoning Tasks', they have discovered that, in science learning at least, it is far from the case. Figure 4.2 shows the proportion of children at the different Piagetian stages in a representative British child population. It indicates that something like 5 or 10 per cent of children are still thinking intuitively at the age of 12 or so, and the less able child is likely to be one of these. Shayer and Adey press the point strongly that the ages of attainment of each of the various stages of cognitive development are significantly higher than classical Piagetian theory suggests. It is worth noting that Shayer and Adey subdivide the last two Piagetian stages into early concrete and late concrete, and early formal and late formal stages. Even by the age of 14, some 20 per cent of children are still at the early concrete stage of cognitive development. It is interesting to speculate that it may be no coincidence that this figure corresponds with the 'one-in-five' of the Warnock population (see p. 24). Going back one stage further, it is possible to hypothesise that less able children in their early teens might remain more egocentrically biased than has hitherto been believed. This is the intuitive stage of cognitive development that traditionally is held to be passed by the age of 8 or so. Yet this kind of thinking, whereby children see everything from their own immediate world view and in relation to themselves, seems to epitomise the child who is unable to conceive that the same word may convey two concepts. The egocentric, intuitive concept of 'evaporation' which Beveridge exemplifies cannot be displaced by the scientific concept, nor can the two concepts be held ambiguously and simultaneously.

It is only during the formal stage of cognitive development that purely abstract thought, such as hypothesising about experimental results, making inferences from data or recognising interdependence between variables, is supposed to be possible. It appears from Shayer and Adey's results that less than one-third of all children will have reached this stage by the time they come to leave school at the age of 16. Similar data have been collected in the United States, where it has been shown that most children in their early teens are functioning at the concrete operational level, as are also a large proportion of college students, and approximately 50 per cent of adults (Vachon and Haney, 1983). If this is so, it indicates a potentially massive mismatch between curricular content and the ability of the majority of children to cope with it. Science curricula whose development have been based on Piagetian principles (e.g. SCISP, 1974) are now open to criticism on the basis of these findings.

Working in collaboration with the SSCR (1984), the Children's Learning in Science Project (CLIS, 1984a,b,c) has produced a series of researches that focus on the confusion in secondary school children's minds about the nature of heat, plant nutrition and energy. At best, 5 per cent of 13-year-old children showed a thorough understanding of latent heat and temperature constancy during change of state for example, and very few understood the problem in terms of the behaviour of particles. Bearing in mind the revised 'ages of stages' of Piagetian development, it is necessary to question whether the abstract concept of the particulate nature of matter, quite unobservable in any direct way, should be on the science syllabus, at least in the junior part of the secondary school. Yet we have seen it taught in many schools. Even by the age of 14, less able children will find it literally inconceivable that air could consist entirely of discrete particles, with nothing between the particles (Osborne and Freyberg, 1985). Thus, in the preparation of science materials for children in their early teens, it is necessary constantly to be aware that children may not be so intellectually advanced as classical Piagetian psychology would suggest.

Shayer and Adey (1981) provide a taxonomy by which a science syllabus can be analysed for the level of intellectual demands it makes on children. The reader is advised at this point to look at figure 4.3 and table 4.1 in order to see how a topic called 'The Transport System' has been written for children at the intuitive/early concrete operational stage of cognitive development. The most important of the concepts embedded in the text are associated with the criteria that, according to Shayer and Adey, characterise the reasoning patterns of children at this particular stage of development. Similar criteria can be produced for all the stages of development recognised by Shayer and Adey, and they also provide a list of criteria relevant to physics and chemistry.

Finally in this section, it is necessary to point out that neo-Piagetian psychology does not require that all concrete, or all formal concepts should be eschewed in the preparation of materials for children thought to be at the intuitive or concrete operational stages respectively. In order for cognitive abilities to be nurtured, and to encourage the acceleration of cognitive development, there must be challenge and the opportunity to move on. It is essential to bear in mind, for instance, that formal concepts such as relative humidity can be approached through concrete experiences. As the child progresses, formal concepts can be taught that are further and further removed from concrete experiences, and progressively more

THE TRANSPORT SYSTEM

In some ways our bodies are like towns or cities. The people in the town need food and water. So these have to be taken to the houses where the people live. The milkman delivers by van, which travels on roads. Water is transported to buildings in large underground pipes. The roads and pipes form part of the transport system of the town.

The human body contains many organs, both inside and outside it. Your legs and your arms are on the outside of your body, and your intestines are on the inside, and all of these are organs. They all need food and water and oxygen to work properly, and they produce waste products which need to be taken away. So, like a town, the body also needs its transport system. The body's transport system contains blood, which travels round the body in small pipes or tubes called arteries and veins. When your legs are running very fast you can feel your heart beating very quickly. This is because it is a pump, and it pumps the blood around the arteries and veins. If you are running fast, then your legs need food and oxygen faster than normal, so the heart pumps the blood carrying these things faster. The blood travels round and round the body, picking up oxygen in the lungs and transporting it to the organs where it is needed. It picks up food from the intestines, and carries away all the waste materials. Each time it passes through the heart, it gets pumped on its way again. You can see all this taking place in the picture.

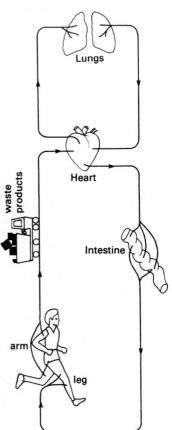

Figure 4.3 The Transport System: an example of a topic written for children at the intuitive/early concrete stage of thinking (adapted from S. Kellington, ed., 1982)

dependent upon previous formal experiences. Figure 4.4 gives an example of this approach in terms of the essentially formal concept of 'relative humidity'.

Cognitive progression can also be encouraged in children by attention to specific aspects of communication. In particular, attention is drawn to figure 5.8 and the associated text, which shows how children's writing can be manipulated to encourage increased objectivity and the move away from intuitive thinking.

Table 4.1 *An analysis of the reasoning patterns required by children reading the text of figure 4.3 for understanding (after Shayer & Adey, 1981)*

Concept in text	Required reasoning pattern
Transport in town is equivalent to transport in body	Interprets phenomena egocentrically in terms of own self (I) Cause and effect only partly structured ('this goes with that') so uses associative reasoning (C)
Water (in towns) is transported in pipes; blood (in body) is transported in pipes	Does not look for contradictions in interpreting descriptive account (I)
You can feel your heart beating	Immediate perception (I)
Legs run fast, heart beats quickly	Relationship limited to one cause and one effect (C)
Heart as a pump	From experience of own pulse (I)
Blood flows to parts of body, including lungs, delivering and collecting	Concrete modelling is the organisation of reality by 1:1 correspondence. Simple comparisons and elementary causes only (C)
Legs, arms and intestines as organs	Can all be seen and felt (e.g. stomach ache), NB not kidney, liver, etc. (C)

I = intuitive, pre-operational
C = early concrete, operational.

PSYCHOLOGY OF NON-VERBAL COMMUNICATION

Before turning our attention in the next chapter to the place of verbal communication in the teaching of the less able, it is worth considering the role that non-verbal behaviours play in the learning environment. These are often not discussed as of real practical importance in the teaching and learning of science, but we take quite the opposite view and present non-verbal communication as part of the context within which verbal communication takes place.

The non-verbal channel of communication is particularly important with less able children because it tends to convey information about emotional states and interpersonal attitudes, and thus addresses itself more to the affective side of the learning environment, which we have argued is of such central importance (chapter 2). Non-verbal behaviours are less easy to control than verbal behaviours, and they are therefore often perceived as being more genuine indicators of attitudes and feelings by children. In particular, it is clusters of non-verbal behaviours that, when taken together, can belie what is being said.

TOPIC	Perceived stage of cognitive development of child	Learning experience			
	Intuitive	concept unattainable			
	Early concrete	sweating when running (C)	→	sweat even more in an anorak (C)	→ hot and sticky (C)
Relative humidity	Mid-concrete	sweating when running (C)	→ sweat even more, running in an anorak (C)	→ steam stays inside anorak (C/F)	→ high steam in, low steam out (F)
	Concrete/ formal	sweat when hot (C)	→ more steam when hot & when wet (as in bath) (C/F)	→ steam=vapour More vapour in second condition than first (F)	→ higher implies relative (NB 2 conditions only) (F)
	Formal	percentage (F)	→ water vapour (F)	→ 'n' variables (F)	→ RH (F)

Figure 4.4 Teaching relative humidity to encourage cognitive progression (concepts can be understood by children at the concrete operational (C) or the formal operational (F) stage of development)

The influence of non-verbal behaviours on verbal behaviours

The value of verbal interactions between teacher and taught cannot be overestimated, and the whole of the next chapter is addressed to this question. It is vitally important to understand, however, that verbal forms of interaction are far from independent of non-verbal forms. The relationship between non-verbal and verbal behaviours is well established in the literature. Abercrombie (1968) has said that 'we speak with our vocal organs, but we converse with our whole bodies'. Indeed, it has been argued that non-verbal communication underpins and moulds the verbal interaction that goes on in classrooms, as we shall see later. Certainly when verbal and non-verbal behaviours contradict one another, children become confused. Inexperienced teachers, and those in initial teacher training, are particularly prone to this kind of mistake. Making verbal statements aimed at controlling discipline often fall into this category. 'Be quiet at the back', implying a desire for interaction

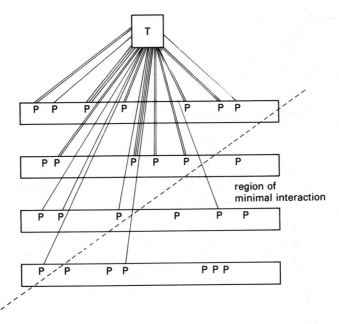

Figure 4.5 Spatial distribution of questions asked by a student science teacher in a 20-minute 'question-answer' session

with those children at the back, can so easily be contradicted by teacher gaze that is almost exclusively concentrated at the front of the class. Not only is gaze concentrated at the front, but so is talk and, in particular, questions to and from children. Figure 4.5 is a record of the verbal interactions over a 20-minute period between a naive student teacher (in her first two weeks of teaching practice) and a class of 14 year olds in a general science lesson, and is typical.

Nor is it true that experience by itself can alleviate this problem (Feidler, 1967). Skills in employing appropriate non-verbal behaviours have to be learned like any other social skill (Argyle, 1975), and this can only happen when the learner is aware of the need for them. Flat, monotonic speech is not only boring in itself, but it is inefficient, because it is not giving sufficient clues to the listeners as how best they should react to it. It can be enhanced by such prosodic signals (see figure 4.1) as placing stress on certain key words in sentences, or by raising the pitch towards the end of a question that requires an answer. For less able children, exaggerated prosodics can be very helpful, especially in science talk, which like science reading, can be turgid and information centred.

Newsreaders on radio and television often have to communicate similarly information-saturated materials, and therefore have similar problems to overcome in terms of holding attention and maximising the flow of information. To attempt to copy their intonation and pitch, keeping a few words behind them, is a salutory experience, and a good indicator of how dull teachers must sometimes sound to children. Kinesic movements associated with speech, such as hand movements, head nods, gaze shifts and facial expression, can all serve to reinforce verbal behaviour. It appears that small bodily movements such as hand movements are seen to relate to small units of speech such as words and phrases, up a hierarchy to where large shifts in body posture, such as shifting the weight from one leg to another, mark changes in paragraphs or longer periods of speech. Children are intuitively capable of decoding such messages, and so become alerted to the fact that a new point of departure in the content of the spoken word has been arrived at. A large number of teacher behaviours will encourage children themselves to talk more. Such non-verbal reinforcers as nodding, smiling, leaning forwards and making 'uh huh' noises are helpful in this respect, and are probably underemployed. So important are non-verbal behaviours in this context that Argyle (1975) argues they may actually be more important than verbal behaviours in encouraging interaction between teacher and taught.

A supreme example of this is given in Rowe (1974a,b), who demonstrated that increased use of silence can, paradoxically, increase the amount of verbal interaction between teacher and taught. When teachers were asked to designate what, in their opinions, were the five best and the five worst pupils in their classes, it was discovered that the five best pupils were given nearly two seconds to respond to a question, but the bottom five were allowed less than one second. In both cases, the length of time that science teachers were prepared to wait for a response was consistently low. On the average, if a child had not responded to a teacher's question within one second of it being asked, the teacher would repeat, rephrase, ask a different question, or call upon another child to answer. When 'wait times' were increased by training to between three and five seconds, there were significant changes in a large number of pupil responses. Amongst these, and of special importance to the slow learner, is the increase in length of the pupil response (see chapter 5, pp. 79–83), the increase in incidence of pupil responses, and the increase in confidence that less able children developed in their own ability (see chapter 2). And all this happens directly as the result of a couple of seconds' non-verbal behaviour on the part of the teacher. It is interesting that, as a result, teacher expectations for the performance of slow

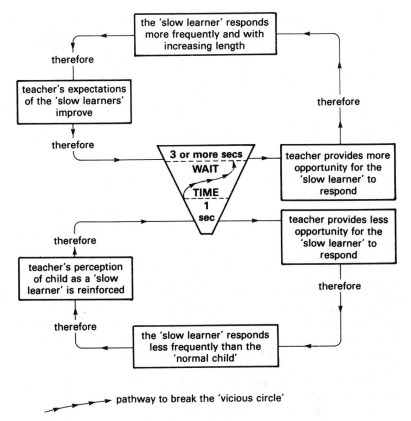

Figure 4.6 The 'vicious circle' and 'wait time'

learners were also shown to be heightened. It appears as if the teachers' non-verbal behaviour towards less able children is influenced by their perception of these children as less able, another example of the oft-quoted 'self-fulfilling prophecy' in education (Rosenthal and Jacobsen, 1968). Figure 4.6 shows how wait time can break this vicious circle of non-response. Sad to relate, however, teachers' expectations are not always that easily modified. There are studies that suggest that, when pupils violate teacher expectations, teachers react to maintain the groups in the expected order (Rubowits and Maehr, 1971, quoted in Rowe, 1974b), which gives credence to the notion that separation of underachievers from other children might sometimes be beneficial. In any case, it is clearly important that teachers react appropriately to pupils who violate their expectations.

Under normal conditions, where teachers had not been alerted to the significance of wait time, not only were less able children given less time to respond, but the verbal responses from the teacher were different from those they gave to the most able five pupils. Whereas the most able children were praised when their answers were correct, there seemed to be no correlation between praise and correctness of answers with less able children. This ambiguity of teacher feedback makes it difficult for less able and underachieving children to evaluate their contribution, reducing its worthwhileness to others as well as affecting the confidence of the children themselves. When wait time was increased, the amount of talk initiated by the bottom five pupils increased tremendously, from approximately 1.5 per cent to 37 per cent overall, although at times even more than this.

The use of classroom and laboratory space and verbal interaction

A teacher's awareness of the value of space as a 'specialised elaboration of culture' (Hall, 1966) is one of the more easily quantifiable of the non-verbal behaviours. According to Hall, human beings may be imagined as being surrounded by a series of more or less concentric psychological bubbles, the closest of which is only about six inches away from us and forms an invisible psychological 'no-go' area. Teachers of science will be acquainted with the inverse square law. It has been suggested that an analogous situation might apply to proxemic behaviour (the way in which space is used), although in this case it is an inverse cube law, $I \propto 1/d^3$. In this case, I represents the influence a teacher is having on the child and, as in the case of the inverse square law, d represents distance, here the distance between teacher and child. It is extremely unlikely that any quantitative validity is attachable to this idea. The concept of I is, for instance, not well defined, and the influence of distance between people is known to be partly culturally dependent. Nevertheless, the formula does serve to alert us as to the possible importance of distance between individuals. Children are often seen to become uncharacteristically diffident when teachers come to sit down next to them. The child may interpret a close encounter of this type as threatening, domineering behaviour, and an invasion of personal space. This is ironic when it is considered that teachers are also making, often intuitively, decisions about their own personal response to the needs of children. When teachers approach close to children, not only are they invading the personal space of others but by the same token they are welcoming those others into their own personal space. So what is perceived by the one as act of intimacy and a show of affiliation is perceived by the other as unwarranted threat. This is

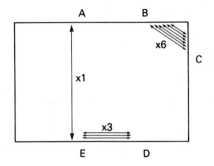

For each episode of verbal interaction between
two people facing each other (A and E) there are
six episodes between two people sitting at
adjacent corners (B and C)

Figure 4.7 Participation by people seated round a square table

particularly likely to be the case for children lacking in confidence
and generally less able to cope, for such children may have come to
associate that kind of proximity with punishment, for at that
distance hands can grip and smack.

Where children choose to sit in a laboratory is also an indicator of
their willingness to relate, and a measure of their affective
commitment to the lesson. Hall (1966) has shown that people
choose to sit round a square table in positions that reflect their
perceptions of others. Figure 4.7 shows how relatively little
co-operation will occur between two children sitting opposite each
other (child A and child E) as compared with two (child B and child
C) sitting adjacently at the corners, where six times as much
co-operative talk may take. place. A semicircle of desks or chairs
encourages discussion amongst children (Richardson, 1967). So the
alert teacher is able to manipulate discussion either specifically
between children by subtle changes in proximity between individuals, or generally by arranging the furniture. It should also be
remembered that, when children are allowed to choose their own
seating positions, these will have something to say about the child
himself. Those consistently seated at the back may indeed be
displaying in their non-verbal behaviours a general uninterested-
ness in work. They may, however, be introverts who do not require
as much affiliative contact with their teachers or with other children.
Again, they may be children who are simply playing 'hard to get',
and who really do want the teacher to make active contact with
them.

The way in which teachers make use of space in the laboratory is so important that it may mould the more overt verbal interactions (Reid, 1980). Verbal interaction patterns in biology classrooms correlated highly with the position of the teacher in the classroom. Teachers who were 'itinerantly involved' (ITV), that is who were moving about the well of the laboratory or classroom, were significantly more likely to receive pupil-initiated questions, lectured less (by almost 50 per cent), were more likely to take up and use pupils' ideas, and were more likely to have pupils who were working constructively on their own than their 'blackboard confined' (BBC) counterparts (figure 4.8). In general, ITV teachers interacted with individuals and with the class more than did BBC teachers. When students in initial training were asked to concentrate on just two aspects of proxemic style – the amount of movement they made around the laboratory and their distance from the blackboard – there was a marked increase in pupil involvement (figure 4.9). This, then, is a useful technique for getting children more actively involved in their science lessons. In addition, it gives

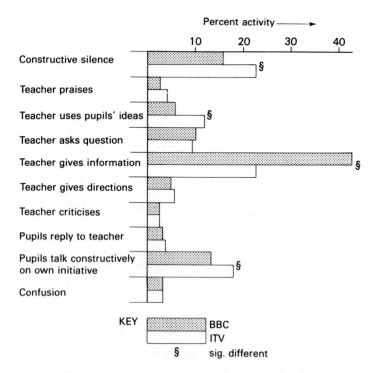

Figure 4.8 Different verbal interactions between blackboard confined and itinerantly involved teachers, showing the effect of teachers' use of space on verbal interactions in the classroom

the teacher increased feedback about what individual children are capable of doing and insight into 'where the children are' (chapter 5), which is, we argue, one of the essential prerequisites to providing the right kind of materials for less able children.

In conclusion, non-verbal communication is important because:

- it is an emotive channel – teachers must be aware of the ability of even the least able and demotivated children to decode leaked information

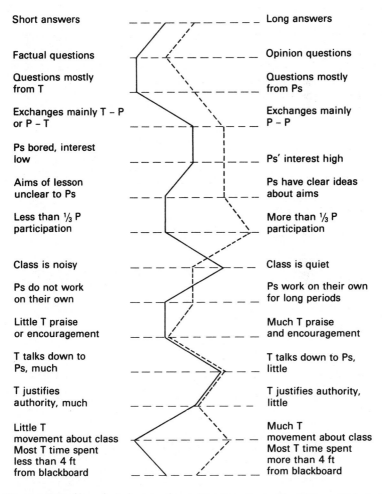

Figure 4.9 Profiles of student teachers' (T) interaction with children (P) before (——) and after (----) they were asked to concentrate on two aspects of proxemic style (teacher movement about the class and distance from blackboard)

- it controls the context in which verbal communication occurs, which means it controls to a large extent what is actually said
- it runs parallel to the verbal channel, and occurs at the same time; its sensitive use therefore increases the efficiency of the communication system as a whole.

Bearing this in mind, it is time to consider the role of verbal communication in the teaching and learning of science.

—5

Communicating in and through science

CHILD, TEACHER AND CONTEXT

The instrumental role of communication in the teaching and learning of science has been discussed in constructivist terms in the last chapter. Teachers who are good communicators are better teachers than those who are not, and we shall discuss how specific aspects of the communication process can be facilitated later in this chapter.

However, communication has another, and probably even more important, role to play. In chapter 2 it was argued that the challenge to the science teacher is essentially the challenge to motivate the child. Motivation is the crucial ingredient that encourages less able children to take an active interest in the world of natural phenomena. It follows that the role of the science teacher is to create an atmosphere in which the child's interest can evolve from conditioned indifference or apathy (because of past failures), to passive acceptance and ultimately to active interest.

The first step in this process is to ensure that the child feels secure in the learning environment. We feel that this is not always achieved in mixed ability classes, if by 'mixed ability' is meant classes containing children with a large range of mental ability and at widely differing stages of cognitive development. As we shall see later (chapter 9, p. 194), there is a current trend to eschew a system that places 'cheek by jowl' (Judge, 1984) children who can and children who cannot. In this situation there is a real danger that less able children are constantly reminded of their weaknesses and failures by comparison with their more able peers each minute of the day, with the effect that they often become conditioned to failure. As a result of this 'learned helplessness' (Seligman, 1975; Dweck, 1984), many children become disillusioned with the education process, perceiving themselves as being rejected by it. In order to counteract this 'loss of dignity' (Hargreaves, 1982), these children often develop their own 'counter culture' (Musgrove, 1979). They derive some sense of satisfaction in rejecting not only science, but possibly schooling in general. Such children frequently present as inscrutable, uninterested, and even unapproachable and deviant in the laboratory (Smith, 1983). Clearly, before science teaching can start to be effective with these children, they have to begin to have some confidence in their own abilities to succeed. Often a complete change of physical environment helps (chapter 9), but where this is not possible then a change in the perceived learning environment is essential. In our opinion, no progress is possible with children who have been so demoralised without major changes in teaching content and method but, above all, in environment.

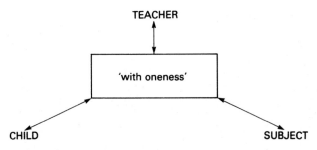

Figure 5.1 Communication as interaction

There is a great deal of precedent both in theory (Foster, 1984; Jenkins, 1973) and in practice (see chapter 9, p. 203), for positive discrimination in favour of the least able, and we cannot believe that such positive discrimination is achieved by requiring such children to work competitively with the more and most able. In 1978, a government report (DES, 1978b) concluded that in mixed ability classes in comprehensive schools teaching was too traditional and pitched at a level that the teacher thought was appropriate to the majority of the class. In this way, both bright children (Freeman, 1979) and the least able suffer. Brennan (1979) maintains that, despite claims for the merits of mixed ability teaching, and with a concern specifically for the curricular needs of slow learners, his project team had 'not observed the needs of slow learners being satisfactorily met within this kind of school organisation', that is, within mixed ability classes with no help for children with learning difficulties. Positive discrimination means special facilities, special organisation, but, above all, special teachers.

By 'special teachers' we do not mean, necessarily, teachers who have been specially trained for working with the underachiever or the disillusioned. Not all secondary schools have such teachers, although if a school is fortunate in this respect their advice will clearly be welcomed. Rather, we mean teachers who have been alerted to, and who believe in, the worth of such children. These are the teachers whose view of the communication process is in accord with the principle of 'with oneness' enunciated in the last chapter, and who perceive communication as the key to interaction between teacher and learner. Caleb Gattegno (1970), puts it like this:

> In this context, teaching becomes a new activity originating within the complex of 'knowing people' who meet deliberately for the explicit purpose of changing time into experience with the greatest efficiency possible.... The relationship of teacher and student is one of a gift of each to the other.

All too frequently, when visiting schools and talking to experienced teachers, we are told that student teachers have little chance of performing well with a particular class, some of whose members are unable to write their own names, much less know one end of a test tube from another!

Of course, interacting with children by communication through science must achieve more tangible outcomes than fine feelings of affiliation and well-being. There are intellectual as well as affective gains to be sought. To help achieve these we must turn our attention to some of the more important individual components of the communication process.

SPOKEN LANGUAGE IN SCIENCE CLASSROOMS

Spoken language as a signpost to where children are

It has already been said (chapter 2) that children do not attend their first science lesson in primary or in secondary school with completely blank minds about science. Within the first few years of life they have developed their own internally consistent meanings for many of the concepts used in everyday science teaching. They know that 'balance' for instance is required to prevent them from falling off their bicycles, and that if proper attention is not given to balance the result can be painfully scraped hands and knees, if not worse. Osborne and Freyberg (1985) believe that these intuitive ideas about science are often strongly held, often significantly different from the views of scientists, and often not well known by their teachers. The reality of the scientific concept of 'balance' comes a poor second to the reality of the experience of 'balance' that the child has had from learning to ride. In this sense, the science classroom comes off worst, for it is 'less exciting, less dramatic and … less dangerous' (Osborne & Freyberg, 1985) than the 'real' thing. Because these intuitively held views are sensible and coherent from the child's point of view, they often remain uninfluenced, or can be influenced in unanticipated ways, by science teaching. Solomon (1983) warns teachers to be on the alert for differences between what she calls 'out of school' knowledge, which is what children bring into school with them, and the 'scientific knowledge' that science teachers must attempt to teach. Science teachers, in their eagerness to 'season their teaching with plenty of examples and analogies from the everyday world', can often imply in so doing that there is no basic difference between science knowledge and social knowledge, and this can confuse children. 'Science', she says, 'is different.'

The facility of children to recognise the subtle differences in context between the way in which a teacher uses a particular word such as 'balance' and their own perceptions of what the word means and to be able to cope with the ambiguities that sometimes exist between 'social' and 'scientific' knowledge is probably dependent upon their ability to cope with abstractions, at the very least an innate ability to perceive natural phenomena in a more objective way than is possible at the 'intuitive' stage of cognitive development. If the neo-Piagetian view of cognitive development is true, then by the time children have reached the age of 11 or 12 they are more than likely capable of working at the concrete operational stage. Even if they are still at the intuitive stage of thinking, as some will be (chapter 4, figure 4.2), the time is probably right to begin the process of weaning them onto a more objective science diet. In the preparation of science materials for underachieving children then, a two-pronged approach is advocated:

- decide whether the child is capable of understanding the proposed material in appropriate scientific terms rather than inappropriate intuitive terms. Nothing succeeds like success – what is the stage of cognitive development?
- discover what the child already knows about a science topic, and the context in which it is understood – what are the alternative frameworks?

The first approach has been discussed in the previous chapter. The second is a little more difficult, but may turn out to be even more relevant to the slow learner. A number of researchers are beginning to establish that there is a consistency in the kind of alternative frameworks that children hold for different scientific concepts. For example, children's intuitively held beliefs about energy tend to focus on its more everyday connotation as a form of vigour ('devote your energies to this', CLIS, 1984c). As far as the concept of 'heat' is concerned, many children 'seem to think that the sensations of both "hotness" and "coldness" are due to something leaving the hot or cold object and entering the body' (CLIS, 1984a). Concepts of plant nutrition are held in terms of plants obtaining their food from the environment, rather than manufacturing it internally, and that 'food' for plants is anything taken in from the outside (CLIS, 1984b). Driver (1983) shows how children expect to find a constant force acting upon a trolley to result in the trolley moving with constant speed. Watts and Zylberstajn (1981) found that children expect a force always to act in the direction of motion.

Being aware of such possible alternative frameworks is, however, no substitute for being aware of specific frameworks held by individual children. In order to discover these, opportunities need to be provided that will allow the children to discuss between themselves what is their current understanding of a particular concept. This will place an even greater reliance on the heuristic use of language by children (see following section), which in turn will only be useful if the teacher has asked the right questions, and makes the effort to listen to the answers. Listening would not appear to be the most developed of the communication skills demanded of teachers!

The heuristic value of language

Much of this section is based upon observations made in science classrooms and interviews with individual teachers and children. In essence, classroom observation techniques fall into two categories –those employing systematic procedures of recording language, and those using a more naturalistic approach. Systematic analysis of classroom language depends upon the breakdown of all that is said into a number of categories. Probably the best known of these systems is that devised originally by Flanders (in Amidon & Hough, 1967). His early ten-point scale is given in table 5.1. Other systems include Eggleston, Galton and Jones' 'Science Teachers' Observation Schedule' (1976).

Such crude measures of the verbal 'interaction' patterns between teacher and taught have been criticised (see, for instance, Edwards and Furlong, 1978), but from this kind of approach has emerged a series of internally consistent and educationally significant findings. Included amongst these is the now famous 'rule of two-thirds', which states that someone is talking in the average classroom for about two-thirds of the time. For two-thirds of that time, the speaker is likely to be the teacher, and for two-thirds of the time the teacher is speaking the talk is likely to be concerned with the giving of information. This rule varies slightly with, for example, 'better' and 'poorer' physics teachers (Pankratz, 1967) and the kind of science being taught (Reid, 1980). Nevertheless, it provides empirical

Table 5.1 *Summary of the ten categories of verbal interaction used to quantify verbal interaction patterns in classrooms and laboratories (after Amidon & Hough, 1967)*

Mainly talk coming from	Category	Type of behaviour	Examples of talk in the category
Teacher talk, having the effect of indirectly influencing behaviours	1	Accepts feelings	Teacher accepts or clarifies the feelings of pupils (quite rare; usually concern is more for ideas).
	2	Praises or encourages	Often a single word – 'Good!' Maybe, 'I see, yes, well go on then.'
	3	Accepts or uses ideas of pupils	'That's fine, now if we were to do that…' Often finishes with teacher using his own ideas.
	4	Asks questions	Does so with intent that pupil will answer. Not rhetorical, e.g. 'What do you think you are up to John?
Teacher talk that attempts to influence children directly	5	Gives information	Gives facts or opinions; expresses own ideas. Characteristically the most frequently used category!
	6	Gives directions	Commands or orders with which pupils are expected to comply, e.g. 'Now light your bunsen burners'.
	7	Criticises: justifies authority	'You will do as I tell you.' All teacher talk aimed at changing behaviour to acceptable pattern.
Pupil talk	8	Teacher-initiated pupil talk	Any talk by children in direct response to teacher, most frequently the answer to a direct question.
	9	Pupil-initiated pupil talk	Pupils want to talk, and do so of their own free will; typically they ask a question of the teacher.
	10	Silence or confusion	Children quietly working on their own, or informally where much talk makes it impossible for observer to interpret correct category.

evidence for the kind of imbalance between teacher and child talk that so concerns HMI. This imbalance is even worse than at first might appear. It does not necessarily follow that the one-third of classroom talk not allocated to teacher talk is allocated to talking by all or even a majority of the pupils. It may be given to individual pupil talk in the public arena as the result of teacher questioning. In a class of 25 pupils, the opportunity for any one child to talk meaningfully about science is usually very short.

In the UK, Douglas Barnes (1969, 1976) has been at the forefront of the examination of classroom talk by more naturalistic methods. He has recorded entire conversations, transcribed them and analysed them in great detail. This approach can be criticised because of its dependence for success upon the flair of the researcher to recognise patterns in the spoken language, and it is susceptible to both subjective and idiosyncratic interpretation by the observer. There is also a danger that an observer might place too much emphasis on extracts of conversation taken out of context. Nevertheless, the technique is currently proving of great value in highlighting some of the underlying reasons for children's learning problems in science.

Psycholinguists appear to be convinced that the very act of talking *per se* is valuable in the learning process. It is, argues Vygotsky (1962), the very struggle to convert half-formed ideas into articulated speech that crystallises thought. 'I have forgotten the word I intended to say, and my thought, unembodied, returns to the realm of shadows.' Most teachers quickly learn that those scientific principles they thought they understood from childhood take on a new meaning when they have to be converted into language that others can follow, and they realise that they never did properly understand. Inexperienced teachers in particular are prone to use patterns of words designed to satisfy their own need for understanding, rather than listening to their pupils in order to discover what form of language would best suit their pupils' learning needs. Barnes (1969) describes this as 'the teachers' need for a climactic explicitness before he can leave the sequence'.

Giving children the opportunity to think aloud in classrooms is difficult because of both the amount of time it requires and the effect it has on other children, for such talk says Barnes (1976) 'is usually marked by frequent hesitations, rephrasings, false starts and changes of direction'. This underlines the inadequacy of didactic teaching methods (chapter 6, pp. 106–8). The use of words by individuals in classrooms to investigate and explore the relation-ships between their own and other people's ideas or events (such as the results of a practical experiment) we refer to as the 'heuristic use of language', and there are a number of legitimate ways by which it can be encouraged in less able children, without detriment to the attention paid by other children. One of these methods, using the children themselves as teachers, is discussed in chapter 8 (pp. 177–80) and exemplified in chapter 9 (pp. 212–13).

One of the most convincing conversations between two children that demonstrates the value of a heuristic use of language by children learning science is supplied by Barnes in his book *From Communication to Curriculum* (1976). Two boys, Glyn and Steve, are working together on a very simple experiment concerned with air pressure, involving the sucking up of milk from a beaker using a transparent straw. Steve is the 'less able' of the two boys, and it is he who suddenly returns to talking about the experiment.

> S. 'What about this glass of milk though, Glyn?'
> G. 'Well that's 'cause you make a vacuum in your mouth ...'
> S. 'When you drink the milk you see ... you ...'
> G. 'Right! ... You make a vacuum there, right?'
> S. 'Yes well you make a vacuum in the ... er ... transparent straw.'
> G. 'Yes.'
> S. 'Carry on.'
> G. 'And the er air pressure outside forces it down, there's no pressure inside to force it back up again so ...'
> S. 'O.K.'

The interesting thing about this conversation is that both boys come to an improved understanding of the role of air pressure in the process. Glyn, the brighter of the two, is forced to articulate his thoughts more clearly than he would otherwise have needed to do because of Steve's insistence upon explicit answers. Barnes is actually making the point that it is the social interaction between pupil and pupil that has led to such a high level of interaction, although here we are more concerned with stressing the psychological aspect of the heuristic use of language. But it is important to realise that psychological processes cannot take place if the social conditions do not allow, and it is the task of the individual teacher to manipulate the classroom environs appropriately. This is much more than simply a matter of selecting the teaching method appropriate to the child and the content. It is more, as the title of this chapter implies, a matter of general sensitivity by individual science teachers to the whole of the teaching and learning environment, within which specific teaching methods will be selected to procure specified learning ends.

Talking, then, is a route to learning, and it follows that teachers have to learn to control their natural wish to challenge inappropriate explanation from children too quickly. By giving children time to talk, teachers are not only actively listening, thereby informing themselves of the intuitive framework within which a particular child is thinking, but also giving the child the opportunity to come to an improved understanding of the subject material. Science tends not to encourage talk that describes feelings or events; but such narrative talk is an alternative to the more official, formal or 'transactional' language of science, and underachieving children find it easier to communicate by story and anecdote than they do by objective reporting (Perera, 1984). There is an important element of surprise to be celebrated when a child suddenly feels what it is like when the hydrochloric acid vigorously fizzes as bicarbonate is added to it, or when the chimpanzee at the zoo spits at a visitor. Opportunities should be found whereby children, for example, make a public case for fluoridisation of public water supplies, or explain to others how an experiment might be organised in everyday language that carries the essence of the science every bit as accurately as does formal language, but also permits personal feelings to be expressed.

READING MATERIALS FOR THE LESS ABLE

Reading skills in science are crucial to the notion of scientific literacy (chapter 1), and the written word remains a major access route to scientific knowledge and opinion. Lunzer and Gardner (1979) found that children in their first year of secondary schooling spent only 9 per cent of their science lessons reading, and this had increased by only 1 per cent in year four. Not only that, but some 35–75 per cent of the reading that was done was reading from the blackboard or an exercise book.

Readability theory

The advent of mixed ability teaching in comprehensive schools has been accompanied by the concept of 'individualisation'. It is now relatively rare to find a single science textbook distributed to such classes; instead it has been replaced by the ubiquitous worksheet. Today in the UK there has been a recent reversal of this trend due to a hardening of copyright attitudes amongst publishers; many worksheets have had to be destroyed because of threatened legal action.

The application of readability theory to materials written for children helps in the effort to match reader and material. Traditionally, readability has been described in terms of the success a reader has with reading material, although it has long been recognised that there are a number of interacting factors both within the text itself and between text and reader that affect its 'understandability'. These variables are summarised in figure 5.2. Over 150 of such variables

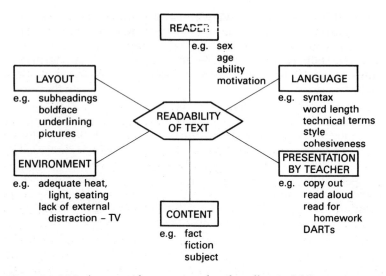

Figure 5.2 Main factors, with some examples, that affect readability

have been recognised, although Klare (1974) claims that the difficulty of reading material can largely be accounted for in terms of just its sentence length and word complexity.

The readability formulae

A host of formulae (Klare has identified over 60) is available for making quantitative measures of readability. The scores so produced can be compared with both the chronological and the reading ages of the children, and any necessary matching can then take place. Thus less able children of, say, 14 with reading ages of, say, 9 can be provided with materials written at a readability level of 9 or 10. The process of measuring the readability level of science literature is lengthy when done by hand, but a number of computer programs are now available (e.g. Reid, 1984a) that accelerate the process. A recent analysis of science worksheets written by science teachers and currently in use in schools (Reid, 1984b) suggests that teachers are intuitively adept at pitching their writing to the reading levels of the target population. Unfortunately, pressures of time and facilities available make it much less likely that sufficient attention is paid to providing the wealth of range of materials so vital for mixed ability teaching, and Reid discovered a disappointingly low percentage of worksheets written for use in conjunction with extension materials for the more or less able. From their observations, HMI comment:

Despite the considerable effort that has gone into the production of worksheets in some schools, styles of teaching rarely pay sufficient regard to the individual needs and abilities of pupils ... pupils of lower ability perform badly and achieve little understanding because the concepts are too difficult for them, while the pace and rigour of the work are insufficiently demanding for the most able. (DES, 1984)

The consensus of opinion (e.g. Perera, 1980) seems to be that the application of the formulae is a useful first step in the selection of graded material for children's learning. However, Reid (1984b) lists a number of reasons why they are inadequate *per se* in solving the problem of children's reading. For example, the formulae are unable to distinguish between polysyllabic words and their conceptual difficulty. 'Aluminium', with five syllables, is much easier to conceptualise than the chemical concept of the 'mole', with only one syllable. The formulae also fail to allow for the fact that the repetition of a polysyllabic word in several different contexts might actually improve a child's chances of understanding it, thus reducing rather than increasing its readability. Great stress has been laid on the fact that many scientific words have different connotations in everyday language, and although often monosyllabic (heat, light, force) again they add to the difficulty of the passage. Cassells (1980) has shown that even such everyday words as 'initial', 'tabulate' and 'contrast' are not understood by 25 per cent of pupils in their fifth year of secondary schooling. Nor can the formulae take account of style. Sometimes long sentences are easier to understand than short sentences. Cassells and Johnstone (1983) noted that internally contradictory terms, for example 'most dilute', and complex sentences with embedded clauses all contribute to depress children's understanding.

Worksheets often make the problem of understanding more acute. Too heavy a commitment to the readability formulae can result in artificially short sentences, and too great a restriction on the use of technical terms. There is a tendency to argue that, because certain children are poor readers, the demands put upon them should be decreased. The danger is that such demands are decreased to such a point that the child is not sufficiently stretched, so that his or her reading becomes even more retarded. Some of the principles of good worksheet production are discussed in the ASE (1980) document, Study Series No. 16, *Language in Science*.

The role of pictures in readability

The readability formulae also fail to take into account the role of a large variety of illustration in making science more understandable (Reid and Miller, 1980). When pictures are used to illustrate science

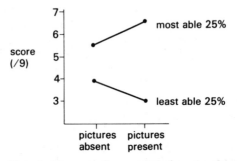

Figure 5.3 The effect of pictures on the memory of most and least able children learning a science topic (after Reid & Beveridge, 1986)

writing, its readability level is lowered (Reid, 1984b). This in itself is evidence of interaction between picture and text in the mind of the writer, if not the reader, and may go some way to explain the so-called 'picture superiority effect' (Reid, 1984c) in science learning. Surprisingly perhaps, there is no evidence that pictures work by indirectly improving the motivation of the learner to learn (Bock, 1983; Reid and Beveridge, 1986). Rather, the likelihood is that learning is enhanced only when information is reinforced by being present in both picture and text form (Levie and Lentz, 1982). What is more, pictures appear to exert an influence on certain kinds of science learning rather than upon others. Reid, Briggs and Beveridge (1983) and others have shown a picture effect on memory facilitation for science materials, but no improvement in understanding of implied meaning or embedded concepts. Again, this is not an overall effect. Samuels (1970) has long been distrustful of the effect of pictures in helping children to read, for example, and has been able to demonstrate where they distract from the learning process. Reid and Beveridge (1986) have demonstrated that pictures can enhance more able children's learning of science, but that the same pictures actually distract less able children (see figure 5.3). It is argued that less able children, finding the text difficult, have their attention drawn to the pictures, which are less rich in information. Certainly, in an experiment with over 200 13 year olds, Reid and Beveridge (1986) were able to demonstrate that children learning from materials containing pictures spent no more time overall on the learning exercise than children learning from identical text containing no pictures, and that on average 20 per cent less time was spent on the text when pictures were available.

It may well be that the composition of the picture plays an important role in its efficacy as a learning aid. Reid, Beveridge and Wakefield (1986) showed that where scientific pictures were inappropriately coloured, or where figure-ground clarity was low, the ability of 13 year olds to perceive items of scientific interest was significantly reduced. Other researchers indicate that specific colour combinations help perception (yellow on blue, yellow on black, for example), whilst others (yellow on green) reduce observation scores (Wicks, 1986). Broadbridge (1986) has shown that less able children are five times more likely to observe items of interest when they are depicted in the foreground rather than the background of a picture. Barlex and Carré (1985) give many examples of how representational pictures and more symbolic illustrations may be used in the production of science materials. It is vitally important, however, not to take too simplistic a view of the efficacy of this form of communication with less able children, for under certain conditions pictures can do more harm than good in helping them to learn.

Active reading

The point has already been made that the degree of success a child will have with a piece of reading material will depend not only upon a number of different variables, such as those summarised in figure 5.3, but also on the way in which some of these variables interact with each other. In particular, recent research has shown that interaction between the reader and the text itself is of major importance. The language of written science is inherently different from that of most other subjects, and children, especially the less able, need help in coming to terms with it. Davies and Greene (1984), in their excellent book *Reading for Learning in the Sciences*, itself based on a Schools Council project (Lunzer and Gardner, 1979), lay stress on the importance of actively involving children in the reading process. Because the language of science is expository rather than narrative, it tends to be more turgid, more information oriented and more succinct than the reading materials that children are more used to.

Recently some attempts have been made to increase the narrative flow of some science reading materials (Kellington, 1982). An article on 'acids in the kitchen' gives the flavour of this new approach.

'If you get an upset stomach caused by acids in your stomach then bicarbonate (bye-car-bon-ate) of soda makes you feel better. That's because it's an alkali – it reacts with the acid.'
'Is that the same as Milk of Magnesia?' asked Raj. 'It cures an upset stomach too.'
'That's right Raj,' said Nita, 'it's an alkali as well.'
'Have you got any Milk of Magnesia, Gran?' asked Raj.
'Milk of Magnesia!' said Gran. 'We'll all be needing that – I have just

overcooked the dinner because of all these science experiments. Now out, the two of you!' (*Reading About Science*, Book 4)

In this series, the children are given some useful advice on how best they should approach their reading sessions:

> To learn about science from these books you will have to do more than just 'follow the words'. It will help you to learn if you
> 1. link what you read with the science you already know
> 2. link what you read with things you know from everyday life
> 3. ask yourself questions about what you are reading and then try to work out the answers
> 4. ask yourself afterwards what you have learned from your reading. (Kellington, 1982)

These general instructions are then translated into specific tasks, all of which are aimed at getting the child to read reflectively.

Such active reading in science needs careful teaching to children who have already become conditioned to passive scanning of pages of narrative text of the story-book type. In most science reading material, each sentence contains information that either is supposed to be immediately relevant or is building up an argument for a future climax. Unless children have prodigious memories, and those who form the subjects of this book do not, it is very easy for them to lose track of an argument because they have forgotten a key point in the development of that argument. Techniques are required, therefore, that encourage them to stop periodically and reflect on what they have read so far, and to try to make sense of it. If this can be done at optimal times in the text to suit natural pauses in the development of an argument it is likely that learning will be enhanced.

Active reading means reading for specific purposes, and these purposes have to be made clear to the child. One of the ramifications of this for teachers of underachievers is that they themselves need to have intimate acquaintance with the materials they are providing for their pupils. The specific nature of the instructions is the key to ensuring active, reflective reading by children. Thus, rather than telling a child to 'read through the first part of the worksheet', or 'read pages 61 to 65 and pick out the main points', the teacher might say 'underline in red all the words that refer to the parts of the electric bell'. Such 'directed activities related to text' are referred to by the acronym 'DARTs'. Davies and Greene (1984), in chapter 3 of their book, not only list some 20 or so DARTs, but give detailed exemplars of how they might be used and the different kinds of science materials for which they are most appropriate.

DARTs represent an important technique in the teaching of less able children, and individual teachers will enjoy the challenge of developing new techniques for themselves. Children might be asked for instance to complete certain words for themselves (a form of cloze procedure whereby every *n*th word is deleted from a passage and replaced by a line of constant length), to cut up a scrambled text with scissors and stick the sentences back in their files in an order that makes sense; this could also be done with a series of scrambled diagrams. They might be asked to complete a partially completed table, the degree of completion having been decided by the teacher, depending upon a particular child's state of knowledge or rate of progress. In this way individualisation is still possible. Other DARTs activities might include children having to decide on a title to a piece of written material, perhaps after discussion with their peers; cutting out labels and attaching them to the relevant structures on a diagram; completing an unfinished diagram; making up lists of questions to act as a test for their friends; or supplying their own examples of the application of certain scientific concepts (say gravity or density) to everyday life. This last is a particularly important aspect of working with slow learners. Figure 5.4 shows how some DARTs activities have been applied to the passage on 'The Transport System' given in figure 4.3.

The ease with which children can relate science to everyday life is often seen as paramount in determining their interest in it. A number of publications recognise this phenomenon. For example, the *Science at Work* project (Taylor, 1979) provides excellent resource materials on such topics as 'making cosmetics', 'soap', and 'perfumes and essences'. Nevertheless, even such worthy efforts to relate the concepts of science to the real world of the teenager are not always successful. It seems to be that the overall context is what is really of importance in this respect. Chapter 9 presents a number of case studies of successful science teaching to the less able, and one of these, the alternative curriculum strategy (ACS), demonstrates how immersion in the real world can often stimulate children.

Another DARTs activity that children seem to enjoy is the making of functional models. Although not in itself a reading activity, it is an activity that is directed by reading. In an account of human reproduction, for example, children can be required to cut out, colour and then stick together a 'functioning' picture of the foetus in the uterus. At different stages in their reading, they will cut out specified parts of the anatomy. Such models are easy to produce (see, for example, figure 5.5), although some are professionally marketed (e.g. Delaney, 1966; Llewellyn-Jones, 1986).

ACTIVE WRITING

Much of what has been said about active reading also applies to

Things to do when you have read the story

1. Working with a friend, try and put the words in the LIST
   below into their proper places in the CHART.

LIST

Roads
Organs
Town
Blood

CHART

	Body
Houses	
	Veins
Van	

2. Cut out the pictures of the three lorries, and stick
   them onto the picture in what you think are the best
   places. One lorry, the one carrying waste products, is
   already on the picture.

oxygen just   blood full of   to body organs
collected   food

3. Underline in red all the words in the passage which help
   to explain the job of the heart.

Figure 5.4 Some DARTS activities applied to the passage on The Transport System
(see figure 4.3 for text)

active writing. Writing is, in one sense, necessarily an active
process, and children can actually be seen to be doing something,
unlike when they are reading. However, it would not do to be lulled
into a false sense of security! In just the same way as teachers should
be careful not always to ask 'closed' questions when talking to a
child, they can encourage writing by asking 'open' questions.
Rather than asking for three characteristics of animals, demanding
three short answers that are either 'right' or 'wrong', and then in a
separate question asking for three characteristics of plants, it might
be better to ask children how they could tell if a monster from
another planet was an animal or a plant.

Whilst less able children often do have immense psychomotor
difficulties with the physical skills demanded of them by writing, it
nevertheless remains a powerful device for helping children to
organise their thoughts. Lunzer and Gardner (1979) found that,

Instructions:
1. Cut out PART A. Glue it in your book.
2. Colour PART B pink. Cut it out and glue it on to PART A.
3. Colour PART C yellow. Cut it out, glue it at point X only.
4. Cut out PART D and glue it on to D.
5. Colour PART E red, leave F white.
6. Cut out E and F in one piece.
7. Glue down Z only, and place E and F on A.
8. Put the heading 'A Tooth', and label the different parts clearly.

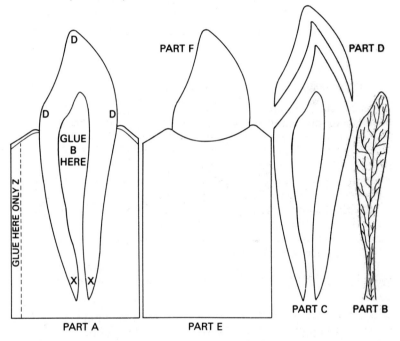

Figure 5.5 A cut-out tooth

typically, a child in year 1 of secondary schooling spends about 11 per cent of the time in science lessons writing, although this has increased to 20 per cent by the time year 4 has been reached. Nevertheless, about half of all writing time is spent copying from the blackboard, or writing dictated notes. Sometimes, entire worksheets are copied by hand into files! Watkins (1981) identifies seven reasons why children should be encouraged to write in science lessons. In this section, we limit attention to just one of these – the significance of active writing as a facilitator of, and an expression of, active thinking.

One advantage that writing has over talking about science is that not so much stress is placed on memory, an important consideration with the less able. Children are able to refer back to what they have written previously. In this way, not only are they reminded of how far they have progressed in their description of an observation, but the chances are also increased that they will perceive new relationships between ideas. Even during the physical act of writing, their minds are able to move from idea to idea, for the act of writing is slower than the act of thinking. Thus thinking can often be facilitated through writing, because writing makes time for prolonged thought, an experience often alien to the underachieving child.

Word processors

The struggle that many children have with writing is characterised by misspellings, grammatical and semantic errors, crossings out, spacing irregularities and so on, as well as by the more obvious depiction of the letters themselves (see figure 2.1). The use of word processors is proving beneficial with some children with learning difficulties, including those with physical handicap. Although there have been some fears that the use of word processors might reduce to superficiality the writing of less able children, current researchers are indicating that this is not so (Pearson & Wilkinson, 1986), and that removal of some of the psychomotor difficulties attached to writing can release creative skills.

Behrmann (1985) lists a number of benefits that word processing gives to the underachiever, the most significant of which appear to be that:

- there is no penalty for revising; it is a relatively painless task to adjust spelling, punctuation, position of paragraphs and so on
- it is easy for children to experiment with writing – if an idea turns out subsequently to be wrong or irrelevant, it can be deleted with no adverse effect on the finished product
- interest in writing is maintained
- writing and editing take less time
- frustration is minimised
- it is easy to produce a perfect copy
- by using a dot matrix rather than a daisy wheel printer, type can be produced in a variety of styles and sizes, a useful facility for visually handicapped children
- computerised spelling checkers are readily available (indeed, children can be encouraged to develop their own dictionaries of scientific terms).

Behrmann (1985) quotes the case of Marc, a 13 year old who, before his introduction to word processing, hated writing anything, for he would choose a wrong verb tense and make so many spelling mistakes that even after one paragraph all desire to continue had gone. Because he made so many errors, his teachers had not the heart to ask a language handicapped child to recopy his work time and again until it was perfect. The word processor made such perfection possible and 'reflected the accomplishment, not the struggle.... . This tool allowed Marc to generate written language in a way that made him feel good about his work and himself'.

Of couse, the QWERTY keyboard is in itself too difficult to handle for some children, although many teachers we have talked to have been agreeably surprised at how easily low achievers learn to master such programs as Wordwise and Edword 2. The Special Education Microelectronic Resources Centres (SEMERC, 1986; see chapter 6, p. 121, for details about this organisation) and the Special Needs Computer Centre (1986a) have produced software that allows writing via Wordwise but with minimal access to the console keyboard. It involves the use of a touch-sensitive pad (called a concept keyboard), connected by ribbon cable to the computer. Children simply touch a word on the pad with one of their fingers and it appears in the relevant position on the VDU. Further details are given in chapter 6 (pp. 121–4) and the Appendix (p. 231). One particularly successful word processing program that can be used in this way is Prompt 3, a big-print processor with the added option of using the concept keyboard to provide one-touch typing of difficult words, and text is generated on the screen as the finger is moved about on its surface. Thus words like 'catalyst', 'mixture', 'reaction', and 'precipitate' can be selected from the touchpad, whilst the simpler connecting words with which children are more familiar, 'this', 'and', 'makes', etc., are typed in directly from the QWERTY keyboard. An earlier version, Prompt 2, has been used successfully in the teaching of science in ordinary schools at CSE level. The Special Needs Computer Centre (1986b) quotes one comprehensive school: 'we have observed a phenomenal use of language as the youngsters using the keyboard communicate with each other and develop their ideas through listening, discussion and problem solving. Whereas previously they were unable to reproduce the information they had learned onto the written page, the Prompt 2 word processor has made this possible.'

Concept mapping

Active writing, which excludes copying or writing from dictation, does not necessarily mean a *lot* of writing, again an important consideration with less able children. Provided that concepts have been judiciously selected to suit the stage of cognitive development of the individual, the manipulation of these concepts, and the exploration of the relationships between them, can be facilitated by a species of active writing known as 'concept mapping'. The theoretical basis of concept mapping may be seen as having its root in the notion of 'meaningful' learning (Ausubel, 1968). This is learning that is accompanied by understanding, encouraged as new ideas become suitably integrated by conscious and explicit action in the child's already existing network of concepts. The alternative to meaningful learning is 'parrot fashion' or rote learning. As is the case with DARTs activities related to reading, concept mapping requires an intimate knowledge by the teacher of the logical and psychological sequencing of the material under study. Novak describes concept mapping as 'a process that involves the identification of concepts in a body of study materials, and the organisation of these concepts into a hierarchical arrangement from the most general, most inclusive concept to the least general, most specific concept'. Linking any two concepts will be a 'logical connective'. It is important that the child is explicit about these logical connectives (Novak, Gowin and Johansen, 1983), for they help to make a functional relationship between two concepts. That is, they describe the context that cements the two ideas together. Often they are simple words in themselves, like 'in' or 'by'. Figure 5.6 gives an example of the simplest kind of concept map.

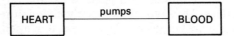

Figure 5.6 A simple concept map, showing how two concepts are joined by a single functional connective

There is a limit to the number of concepts a child can be expected to handle on a single map, and a rule of thumb sets this number at about 20 (Malone and Dekkers, 1984). For less able children it would be fewer, although we know of no research that is available to suggest the precise number. Even six well-chosen concepts would convey the broad organisational structure of a learning programme. It would be feasible for the teacher to produce a simple concept map, maybe with a few of the boxes left empty. The children would discuss the contents, and then write them into the appropriate boxes or shapes. New material presented as an adjunct to the already understood network would require the children to extract the major concepts – first, they would have to recognise them; second, they would have to incorporate them into the already

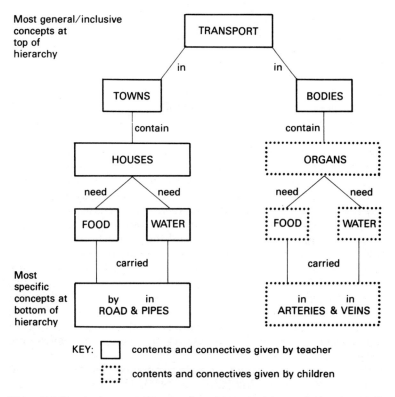

Figure 5.7 Developing a concept map (based on material presented in figure 4.3)

existing map; and, third, they would have to search out the correct logical connectives. The size of the chunk of material to be learned would be predetermined by the teacher in the light of the number of new concepts it contained. Teaching material based on analogy or metaphor, taken from common experience, is particularly suscepti-ble to this kind of approach. In the case of 'The Transport System' (figure 4.3), the teacher might develop the concept map in terms of the analogy, in this case 'towns', and require the children to mirror the map with one of their own, substituting the 'town' concepts for 'body' concepts (figure 5.7). Later, they might be expected to develop the theme by adding concepts like 'arms, legs and intestines', 'blood' and 'heart'.

Styles of writing

Whilst concept mapping requires a minimal amount of writing combined with a great deal of thought, most writing exercises are more prolonged. There are also different styles of writing. The 'Writing and learning across the curriculum' team (Martin, 1976) identified three main kinds of writing along a continuum:

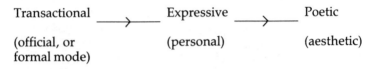

Transactional	Expressive	Poetic
(official, or formal mode)	(personal)	(aesthetic)

They found that, in science, 92 per cent of writing was in the transactional, formal mode, and none in the expressive or poetic modes (the other 8 per cent was written in other 'miscellaneous' modes). There is an obvious analogy here between the expository and narrative styles of writing for reading, discussed above (pp. 87–9). Just as less able children find the expository style of writing the most difficult to read, so they find the transactional style most difficult to write. They find the use of personal pronouns difficult to escape from, and become anxious at their inability to find the 'appropriate' style. A shift to an expressive mode of writing has been advocated by many science educators interested in the communication of scientific ideas by children (e.g. Carre, 1981); the style is more like talking, more familiar to the child, and a better vehicle for expressing doubts, asking questions, and developing ideas.

There appear to be about four main parameters that can be used to describe a piece of writing (Christie, 1986). They can be called the 'content', 'structure', 'audience', and 'purpose' parameters (see figure 5.8).

In broad terms, the *content* parameter focusses attention on the intellectual nature of the exercise. Thus it may be a piece of abstract writing, concerned, say, with the particulate nature of matter; or concrete writing, with the description of an observed phenomenon; or imaginary writing, perhaps a science fiction story of a child reduced to cell size, and making a journey around the circulatory system in a miniature submarine; or, finally, a direct experience, where children might write their impressions of a visit to a nuclear power station or a sewage farm. The implication is clearly that within this, as within other parameters, some forms of writing are inherently more difficult than others.

The *structure* parameter refers to the style of writing – whether, for example, it develops an argument; or whether it is largely transactional or formal; or whether it is narrative or expressive.

The *audience* parameter is another important feature of writing that prescribes its difficulty. Writing for an abstract audience, such as the readers of a scientific journal, the mode aped by most school

PARAMETER	LEVEL	
	Relatively easy	Relatively difficult
Content	'Direct experience'. Choice-chamber experiment. One independent variable (humidity); one dependent variable (numbers).	'Abstract'. Extract from textbook or from scientific paper. Multivariate.
Structure	Expressive; informal use of pronouns OK. Feelings and doubts about accuracy encouraged.	Transactional; formal; typical 'objective scientific stance'.
Audience	Known, shared and liked. An absent classmate, who needs to be informed.	Unknown, distant and abstract. 'Scientists', examiners, etc.
Purpose	Analysis Detecting organisation, structure or pattern in a piece of science. Distinguishing relevant from irrelevant material. Identifying strength of a conclusion.	

Figure 5.8 Classification of the context parameters that prescribe the difficulty of written materials

science writing, is impersonal, abstract and transactional. Clearly such a combination of features of writing is most inappropriate for children with learning difficulties. But just as a more appropriate structure can be selected for less able children, so can a more appropriate audience. The audience can range along an affiliative continuum for example. At one end would be 'self'. Here the child is writing for himself, perhaps in a science diary ('I gave the hamster some grass, but he liked the carrot best'). One step further along the continuum is a friend or a close relative. Here a letter to Auntie Jane, explaining why the iron must be adequately earthed, would give the child some experience of objectivity. Further still along the continuum is the generic audience. Thus, the reader of the school magazine or the local newspaper might be interested to read an account of Halley's comet, or some item of topical interest. Carré (1981) suggests a murder story.

Provide a story about a murder victim which has attached to its body a warning, written in black ink. After the murder the pen was dropped. The ink was matched. Three 'suspects' in the class provide samples of ink. With filter paper and water the class do the experiment to find out who was the murderer. The class write an account of circumstances, evidence, technique and conclusion to the case for the local newspaper.

Widening the audience to a national or international level encourages even greater opportunity for objectivity, by writing television scripts and news programmes, or perhaps part of a documentary to people in the third world, explaining the importance of birth control, irrigation, clean water, and so on. Only at the far end of the continuum would writing for readers of scientific journals be relevant; this provides the abstract audience.

The fourth parameter is the *purpose* for which the writing is designed. Purpose may reside in creative writing, social writing, descriptive writing, analytical writing or logical argument. Once again, it is apparent that some of these are inherently more difficult than others. Creative writing is unusual in science, and demands a personal involvement and imagination, although not necessarily at a high level. A child stranded on a desert island, and with limited materials, might be asked to describe in his diary how he made fresh water when there was no source on the island. Social writing – perhaps explaining in writing to a deaf child an item seen on 'Tomorrow's World', or the social and ethical issues involved in smoking in public – provides another purposeful outlet for children's writing.

The number of permutations between these four parameters and the levels contained in them is high, and an ability to make use of these permutations to grade the difficulty of writing tasks is important. There may come a time, for instance, when the teacher feels that a less able child should try some analytical writing. Traditionally, such writing has taken place in the context of a transactional, highly formal style, containing abstract concepts and written for a distant, abstract audience. For a less able child on this occasion, the purpose parameter, 'analysis' in this case, would be immutable. But the other three parameters could be modified to ease the cognitive load. There is no reason, unless tradition be reason, why an analysis should not be made in an expressive style, using direct experience and a shared audience. As a less able child progresses, the levels within the four parameters can be modified accordingly.

Marking

It is not only the children who will engage in 'active writing'. One of the functions of written work is to provide feedback for the teacher. Adequate marking not only informs children of their progress and points out where improvements should be made, but it also encourages children to feel that the teacher is taking an interest in them. Again, we feel it is of paramount importance to stress the significance of this latter point in the light of the philosophy that underpins this book. If the teachers understand the significance of the writing parameters outlined above, they will not necessarily feel the need to amend a child's description of the action of hydrochloric acid on zinc as 'a seething frothing turmoil' by putting a red line through it and writing instead the word 'effervescence'. There are occasions when the correct jargon is required of course, but there are also many occasions, some of which will have been specially engineered, when this will not be the case.

We are well aware of the time constrictions on teachers. Benton (1981) quotes a letter from an English teacher to the *Times Educational Supplement*, in which the claim is made that the teacher will teach something like 170–200 different children each week. In order to spend five minutes reading and marking the week's work of each pupil, some 15 hours are needed. This works out to two–three hours per night for five or six nights per week! There is no reason to suppose that the science teacher's load could be very much less than this. To spend this amount of time in marking would be not only an extremely debilitating process, but also a misappropriation of teacher time. The message coming over strongly in this book is that the successful teachers of the less able need to know more psychology, need to be more creative in their search for relevant and satisfactory materials and experiences, and need to take more painstaking care over individual record keeping and progress than successful teachers of the most able. They often have to take on these responsibilities without the benefit of departmental resource backing, in the early stages at least. Add to that the drain on their energies from the attention of often socially deprived children and the inevitable failures that will surface from time to time, and one begins to get some perspective of the formidable nature of the task. Whilst it does not make sense to spoil such good work by inappropriate or insensitive marking, it should be remembered that there are viable alternatives. Many children prefer to have their work discussed on an individual basis (Watts, 1980, in Sutton, 1981), and there are opportunities in less formal classrooms where this is possible. This does not obviate entirely the need for a written approach to marking, whereby some permanent record of guidance is provided, but it does reduce it.

—6

A re-examination of teaching methods

INTRODUCTION

We have seen from the introduction to part II of this book (p. 41), that the classical working model used by many curriculum developers is one that incorporates a number of more or less discrete components, and indeed the chapters in part II of this book for the most part reflect these components. Figure 6.1 reminds us that content is necessarily consequent upon aims, and implies that the content itself will to some extent at least control the methods to be employed in the teaching of that content, but so too will the aims and objectives. This is an important idea since it reinforces once again the notion that it is not possible to isolate any one component of the curricular process, for each component interacts to a greater or lesser degree with each of the others.

Thus the methods employed by teachers of underachievers will be influenced in the first place by the overriding aims we have for the science education of these children. For example, a curriculum such as the O level Nuffield physics course, whose philosophical

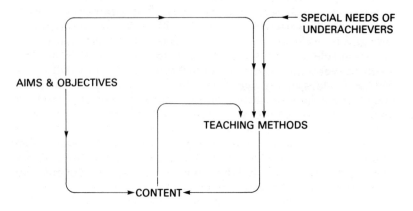

Figure 6.1 Revised components of curriculum model

orientation is towards 'becoming a scientist for the day' will need to employ teaching methods that will allow children to make some discoveries for themselves. This in turn means that they will need to be provided with appropriate laboratory schedules and guides, that they will need to be given opportunities to make decisions about hypothesis making, and that they will require practice in the design of apparatus, some of it intricate, delicate and expensive. Reliance, by teachers and pupils alike, on the help of an experienced technician is another corollary of such a philosophy.

To take another example, typical of many junior science lessons, it may well be that a decision has been taken to make 'improvements in the social development of children' a major curricular aim. A. V. Kelly (1975) is adamant that 'The evidence points clearly towards improved pupil–pupil and teacher–pupil relationships in unstreamed schools, along with a greater involvement of all pupils in the work and the life of the school'. We are by no means convinced of the truth of this statement, either on the basis of the evidence that Kelly uses to come to his conclusion, or with hindsight on the basis of the results that unstreamed teaching have produced in the decade or so since Kelly wrote these words. Nevertheless, improved pupil–pupil interaction is a perfectly laudable and legitimate aim of the science curriculum, and one that we would certainly espouse in the context of underachieving children. Science teaching methodologies are today replete with the jargon of unstreamed teaching techniques, and words and phrases like 'individualisation', 'mixed ability', 'worksheets', 'circus experiments', 'banding' and, latterly, 'profiling' and 'criterion-referencing' are evidence of the pervasive influence of this philosophy.

It is not only the aims and objectives of the curriculum that prescribe teaching methods. In later chapters the idea of negotiated topics is discussed. Without knowing in advance what these might be, it is not possible to suggest whether methods relying on primary source data such as measured light intensity or mass are more appropriate than methods relying on the selection of material from secondary sources such as television programmes or textbooks.

In addition, the special needs and abilities of the children themselves will influence the choice of methods. The population of less able and underachieving pupils is highly heterogeneous in terms of individual learning styles, a state of affairs exacerbated by the fact that differences in the personalities of children also need to be taken into account when considering what teaching methods are most appropriate. There is certainly no reason to believe that general intelligence in itself is any pointer to preferred cognitive style (Witkin, 1975). In other words, our target population is not made any the more homogeneous because it is generally of lower intelligence than the average child in the population, since cognitive style or preferred ways of thinking appear to transcend intelligence. This is reflected in the way in which children respond to experiences of school science. Some children, for example, will be alerted more easily than others to the idiosyncratic nature of specific scientific phenomena – perhaps a unique event that takes place during the course of their normal reading, or during a practical session, and that causes them to question previously held beliefs.

There is a nice experiment that can be performed on the larval or maggot stage of houseflies. It is certainly the kind of experimental work that might appeal to underachievers, for it brings them into direct contact with primary source material, does not need intricate manipulative skills, and generally produces unequivocal results. A number of maggots are placed in the centre of a large piece of card on which sectors have been drawn and numbered. In a darkened room a light is shone from one side of the card, and the released maggots wriggle away from the light (figure 6.2). After a few minutes, the distribution of the maggots in the various sectors is

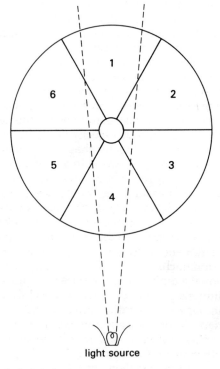

light source

Figure 6.2 Maggot phototaxis

recorded. The experiment is designed to show that the maggots are negatively phototactic. In a class of some 25 children working in pairs there is usually one maverick maggot, at least, that disobeys the rule and moves towards the light, and that will continue to do so however often the experiment is repeated. 'Articulate' or 'field independent' thinkers will find this fact highly relevant, so much so that it is likely to detract attention from the overall or 'global' lesson that maggots are negatively phototactic, and that this is just a manifestation of the overall physiological adaptation of organisms to their environment. The reason for this is that the eggs are normally laid in decaying animal material. Towards the end of the larval stage of the life history the maggots prepare for pupation. It is at this point that they move *towards* the light, near the surface of their habitat. The hard pupal casing is resistant to desiccation, and the newly hatched imago is freed to dry out in the air and fly quickly. Imagoes hatching at the bottom of a dense substrate are liable to damage in their struggle to surface. This impressive biological adaptation often goes unnoticed by the majority of the children, and probably by the majority of teachers, at least to the extent that it is simply passed off as a quirk of behaviour, unimportant in the face of the seemingly overwhelming consistency in the opposite behaviour shown by other maggots. This latter group of children contains within it the 'global' or 'field dependent' thinkers.

It is easier for 'global' thinkers to think of unique events in the context of overarching principles or general patterns. Such children would be able to appreciate the idiosyncratic nature of planetary motion or an expanding universe if first they had been introduced to the principle that could be seen to cement them together. In this case it might well be that the 'big bang' theory would supply a satisfactory contextual 'field'. An American psychologist (Ausubel, 1968) has argued a method of science teaching called the 'advance organiser' in such cases. An advance organiser provides a scientific context or 'field', without itself necessarily adding anything to the main lesson or series of lessons. It is not a summary of the science to come, but provides an opportunity for the globalist to 'subsume' individual items of content within a meaningful context.

'Articulate' thinkers prefer to be more self-directed and are less conforming than global thinkers. This implies that they might become more easily bored with teacher-directed activities than their peers, resulting perhaps in 'deviant' behaviours. Unfortunately, many teachers interpret such behaviour as wilful disobedience, and try to counteract it by controlling ever more narrowly the learning behaviours of these children, thus merely exacerbating the problem. The answer may well lie in giving these children more, rather than less, responsibility for their own learning, a point made by

HMI in one of their *Education Observed* documents (DES, 1985c), and this has important ramifications for teaching methodology.

Such individual differences between children of lower ability puts responsibility for improving communication more squarely onto the shoulders of the teacher, for the children themselves are less able to adjust their behaviours to what the teacher perceives as the 'norm'. It means often that the teacher has to employ extra humility without at the same time surrendering respect for authority. Teachers can be quite easily alerted to some of the more obvious differences between groups of children. It is said that even quite young West Indian children dislike physical contact with teachers. Many Asian children have difficulties in coping in a public classroom with science that has sexual connotations such as menstruation and childbirth. Girls are said to show a greater need for affiliation than boys, and many male teachers can tell of distressing occasions when verbally strong tactics, often successful in the short term with underachieving boys, have produced long-lasting rejection from underachieving girls.

Differences between boys and girls are of special significance in science and technology education. An international study undertaken in 1978 (and reported in Kelly, 1981b) looked at the relative achievement of boys and girls in physics and chemistry in 19 countries around the world. These countries varied considerably in terms of their cultural, political and industrial interests and in terms of the kind of science and technology education received by their young people, yet without exception the boys in each country achieved better results than the girls in the same country, although the girls in some countries, for example, Japan and Hungary, scored better than the boys in some other countries such as the USA and Italy.

A recent project undertaken by Alison Kelly and her co-workers, called 'Girls into Science and Technology' and better known under the acronym GIST, aimed at improving girls' attitudes to science in ten comprehensive schools in three local education authorities in the Greater Manchester area. In spite of a fairly intensive three-year programme, which involved a number of intervention strategies, the attitudes of 2,000 girls to science – as measured by the options chosen in their third year of secondary schooling – seemed to have been little influenced by GIST (Kelly, Whyte and Smail 1984). Reid and Tracey (1985) also found a more rapid deterioration in attitudes to science in the first three years of secondary schooling in girls than in boys.

Sex differences are associated with a number of intellectual activities especially relevant to science, and these include numerical ability, spatial ability and problem solving (MacFarlane Smith, 1964). In each case it is the boys who consistently out-perform the girls. Whether such differences are genetically inherited or culturally induced is a matter of some debate, but the fact that they exist cannot be denied, and the search for methodologies that will lead to an improvement in girls' attitudes to and achievement in science is an important one. In the early days of research in the area it was suspected that girls did not like to be seen competing with boys in a subject area perceived by the girls as 'masculine', the reason being that their own emerging femininity might be seen by boys as being diluted in some way. However, single-sex groups for science, whether in mixed or in single-sex schools, have failed to produce any improvement in girls' attitudes (Harvey, 1985), so it appears that there is no obvious or simple way in which to solve the problem. Certainly the same problem does not exist in biology as it does in physics and chemistry, and science education may find an easier way forward with underachieving children generally by concentrating more on biological content. Head's book *Science through Biology* (1976) attempts to introduce chemical and physical principles through a study of biology, but the book is intended for more able pupils and, in spite of the number of pictures and diagrams it contains, it is unsuitabe for use with underachieving children.

The message of this chapter, so far, is that there is no such thing as a homogeneous group of children whom we can call either 'less able' or 'underachieving'. Such labels are dangerous if they are used to imply that any single teaching method is likely to prove a panacea when applied to these groups. The children in these categories may well differ amongst themselves in even more respects than 'normal' or 'more able' children. The methods employed to teach such children will need to take into account the curricular content and its underlying philosophy as well as any of the more overarching aims of the school itself. The learning styles of the children, their varying personalities, their often reduced capacity for developing working relationships with their teachers and peers (because of sometimes impaired communication skills), and their wide-ranging cultural and genetic origins, including gender, are some of the major contributing factors that need to be taken into account in any discussion on teaching method. If there is one common denominator that helps to identify the target group it is, as we have argued at length, the fact that they are not highly motivated. The disparate underlying reasons for this lack of motivation, some of which are described above, are legion.

A REVIEW OF SOME METHODS IN COMMON USE

If current methods were satisfactory, Warnock (DES, 1978a) would not be talking about one in six children in ordinary secondary comprehensive schools permanently requiring special attention. The next part of this chapter examines some of the teaching methods that have been employed in school science departments over the last century or so, in an effort to elucidate some of the reasons why they have been so popular and yet so characteristically unsuccessful in terms of less able and underachieving children.

The didactic method

Before the middle of the nineteenth century the teaching of science hardly existed in any school in the country. In 1860, for example, only nine schools taught science. Ten years later, 799 schools were teaching the subject (Ministry of Education, 1960). During this ten-year period one of the most significant parliamentary interventions in education took place, an event that was to influence the education process for decades. Even today it is frequently mentioned in the press, on documentary televison programmes and in academia, including the schools themselves, with awe. The Revised Code, introduced in 1862, led directly to the now infamous 'payment by results' era of education, a practice whereby grants to schools, and consequently the teachers' salaries, were made largely dependent upon their pupils' success in examinations. The grant was limited in any case to 12 shillings (60p) per child per annum – 4 shillings (20p) being paid on the basis of a minimum attendance by a child, and the remaining 8 shillings (40p) was paid when the child proved that he or she could pass the examination at a suitable level. Poorly paid teachers (then as now) were forced to rely upon what they perceived as the most efficient method of inculcating science content into their charges. Didactic teaching and rote learning went arm in arm as, by regimentation, rod and repetition, the children and teachers of the day prepared for what was literally to be the day of reckoning. Dickens' schoolmaster Gradgrind epitomises the attitude of the Victorian science teacher: 'Facts alone are wanted in life.' It is also fair to say that didactism was justifiable in those days on more than just the ground of efficiency. The prevailing philosophical view of the child's mind at birth was of a *tabula rasa*, literally an 'erased tablet', upon which had to be imprinted the knowledge and values that the child would require in order both to benefit from and contribute to and conform to society. As the old quatrain has it:

Ram it in, ram it in,
Children's heads are hollow.
Ram it in, ram it in,
Still there's more to follow!

There are of course occasions even in the modern science classroom or laboratory where the didactic approach remains the best method of teaching. However enlightened a science syllabus might be in its concern for process and method, a certain inescapable amount of basic content has to be taught by the teacher and known by the taught. This is especially the case in examination courses, although even here alternative methods can be employed that will fulfil the same aim of maximising the transfer of information. As far as less able and underachieving children are concerned however, few things can be more dogmatically stated than that the didactic method of science teaching is not merely ineffectual, but positively harmful and almost certainly in itself a major contributor to underachievement.

The didactic approach is ineffectual for a number of reasons. One of its assumptions is that information is a valuable commodity *per se*. Information is only valuable, of course, if it is seen by the children as having potential worth. Since less able children will, for the most part, probably not be taking any public examinations at 16+, the overriding need to 'know' is no longer relevant. It is too early yet to predict what role the GCSE examinations will play in the science education of less able and underachieving children, but such syllabuses as are currently available do not appear to us to have any great potential in this respect. Thus, the examination process itself will probably not provide any motivating impetus to the learning of information. The approach also prescribes certain patterns of classroom or laboratory behaviour, namely, a physically and mentally passive attitude on the part of the learner – an ability to sit still and concentrate, to contemplate and take notes. Such disciplined behaviour is often beyond the capacity of less able and underachieving children, either because they do not possess the psychomotor skills for taking notes, let alone the intellectual capacity to understand what it is they are supposed to be taking notes about, or because, as underachievers, they become easily bored with inactivity and so become disruptive.

Didactism is positively harmful in that it reflects a philosophy of teaching that takes the teacher to be the most important person in the classroom. It is the teacher who holds centre stage, who has 50 times as much space to move about in than his pupils (Sommer, 1969), who does upwards of two-thirds of all the talking (Reid, 1980), and who is so often seen as the fount of all knowledge. For 'normal' children this is bad enough, but at least they have the benefit of being able to cope with the knowledge that, although the medicine may be nasty, at least it may ultimately be of some benefit to them. Such a placebo is unavailable to less able and underachieving children.

The prime aim of all teachers should be the demise of their own role. Truly successful teachers are those who, over the years, work actively to ensure that their pupils eventually will no longer need them or any other teacher. These will be the children who leave school secure in the knowledge of their own self-sufficiency and confident that they can not only survive but will enjoy the challenge to survive in a world rich in new and available experiences. Because science and technology are two of the more difficult school subjects, it is natural for teachers of these subjects to want to take a very active role in their teaching. Experienced teachers know where the pitfalls lie, they know the kinds of mistakes that generations of children have made in the past and, having made them, how difficult it is for them to be corrected. There is a natural tendency, then, for the most committed of teachers to want to protect their pupils, and the easiest way to do this is by the didactic approach, which does its best to ensure that the child 'gets it right first time'.

Dictation of notes is symptomatic of this approach. With less able and underachieving children it might be argued that it is even more important that what they do learn is learned accurately, and that didactic methods are even more relevant in this situation. It is a false argument because, as we have already said, it makes the assumption that the teacher knows best. Teachers, curriculum writers and educators generally face a major problem when coming to terms with the teaching of the less able and the underachiever. They are themselves amongst the most able, highly achieving and motivated people within the education system. Everything that is written for and about less able and underachieving children is written by people who have never themselves experienced what it is like neither to know nor to want to know. School in general, and probably science in particular, is perceived by such children as a massive imposition on their freedom. They cannot understand why they are being forced to do things that are hateful to them. They find it almost impossible to relate to people in abstract ways through academic lessons. The didactic approach merely exacerbates their alienation from the system.

The heuristic method

No doubt largely as a reaction to this severely didactic approach, there was a swing of the pendulum at the end of the nineteenth century. The idea was pursued that science was not merely a body of knowledge *per se*, but incorporated a method of working. The body of knowledge that was science was of course rapidly expanding at the time, and the process by which it was able to expand became a focus of interest to science educators. In particular, H. E. Armstrong, Professor of Chemistry at the London Institution in Finsbury, propounded some strong views on the subject. In an article in *The Educational Times* of 1891, called 'The Teaching of Scientific Method' he said:

> I refer to the absence of any proper distinction between the teaching of what is commonly called science, i.e. facts pertaining to science – and the teaching of scientific method. (Van Praagh, 1973)

Armstrong's continuing influence on science education can hardly be overestimated, although the evangelistic overtones that often accompanied his reforming zeal gave rise to considerable criticism. Most often the complaint was that too little time was available to the science curriculum for discovery methods, which for example at times eschewed the use of textbooks. In 1902, Armstrong told the British Association for the Advancement of Science:

> And great care must be exercised that the palate for experimenting, for results, is not spoilt by reading. The use of text-books must be carefully avoided at this stage in order that that which should be elicited by experiment is not previously known and merely demonstrated – a most inferior method from any true educational point of view and of little value as a means of developing thought power. (Van Praagh, 1973)

The claim was never made, however, that all the teaching of science should be by this method. We need to bear in mind, too, that the textbooks of his day were very different from those of today. School biology at the time of the Revised Code in the 1860s, for example, meant largely botany. John Lindley's textbook for use in schools, published in 1866, was called *School Botany, Descriptive Botany and Vegetable Physiology; or, the rudiments of botanical science*. It contains 212 pages of tightly written descriptive prose, under such chapter headings as 'Of thalmaliflora exogens' and 'Of monochlamydeous exogens'. Blaisdell's 1897 *A Practical Physiology*, an American text, was designed 'to furnish a practical manual of the more important facts and principles of physiology and hygiene, which will be adapted to the needs of students in ... normal schools'; but it is very far removed from the kind of practical laboratory guide found in schools today. For instance, in the chapter entitled 'The bones' there are page-long descriptions of the chemical composition of bones, their physical properties, their microscopic structure and 30 pages describing individual and groups of bones. Nine 'experiments' are suggested, the most comprehensive of which is:

> To show how the cancellous structure of bone is able to support a great deal of weight. Have the market-man saw out a cubic inch from

the cancellous tissue of a fresh beef bone and place it on a table with its principal layers upright. Balance a heavy book upon it, and then gradually place upon it various articles and note how many pounds it will support before giving way. (Blaisdell, 1897, p. 55)

It is not difficult to understand why Armstrong was so keen to stress the importance of thinking scientifically through experimentation, and the educational value of the forms and methods of science rather than its content.

There are a number of reasons why the heuristic method did not become more widespread in British science education, and why, by the early years of the twentieth century, what influence Armstrong had had began to wane. Essentially heurism had been overplayed, and was inefficient when taken to extremes. The first world war was a salutory reminder to the British that their scientific expertise was not keeping pace with that of their enemies and, after the war, with that of their competitors. For the next half-century almost, the concern amongst science educators was to introduce the idea of 'science for all' and a 'science of the common man', so that a greater emphasis on content once again became apparent.

The method of 'guided discovery'

It is often claimed that the re-emergence of discovery methods of teaching and learning in science in the 1960s was catalysed by a single event in the year of 1957. Just as the first world war is supposed to have alerted British science educators to the need for a return to the information- or content-oriented approach, so the launching of the first Russian sputnik is supposed to have alerted American science educators to the need for a population of scientists with more flexible and even creative skills. There is no doubt that the atomic age, the space age and the computer age, coming hard on each other's heels, demand a more scientifically and technically informed citizenry than has hitherto been the case. However, the first pressure for curriculum reform was to obtain 'more and better scientific and technical manpower' (Hurd, 1971). The curricular reform in science and technology in the late 1950s and 1960s was aimed in the first place, therefore, at the most able children.

In the UK the most influential curriculum package was undoubtedly that funded by the Nuffield Foundation. The principal teaching method embedded in the various subject syllabuses appears as that of the 'guided discovery' approach to the teaching and learning of science. This, in turn, is based on the theories of such eminent psychologists as Piaget and Bruner. Jerome Bruner focusses early in his book *The Relevance of Education* (1972) on the importance to the learner of the way in which knowledge 'becomes codified and organised'. In effect he makes the point that children's learning is not so much a matter of logical progression, but of progression based on the ability of children to organise their own thoughts in a psychologically satisfactory way. Thus, for example, a basic tenet of Brunerian psychology is that the natural world is too complicated for every phenomenon encountered to be regarded as unique. Were that to be necessary, the immensity of the bombardment of stimuli would very rapidly lead to cognitive overload. In order to cope with this complexity, children learn early on in life to categorise various experiences. They learn to appreciate the underlying order, consistency and pattern in natural phenomena. In this sense, every human being has a natural propensity to classify. For the older child in secondary education, we can utilise this propensity. By a judicious structuring of science lessons, the teacher can 'give the child practice in the skills related to the use of [scientific and technological] information and problem solving'. Children are encouraged to develop their own strategies in the search for psychologically satisfying relationships. Their work is 'guided', often through practical work but by no means solely by such means, through a series of 'discoveries' designed to reveal some kind of ultimate scientific 'truth'.

For instance, the teacher has it in mind that enzymes function optimally at about 37°C. The children are given the opportunity to discover for themselves that salivary amylase catalyses starch most rapidly at this temperature, and that the rate of activity of the enzyme decreases above and below this temperature. They become alerted to the possibility that other enzymes might react in a similar way, and might be guided towards working with a yeast suspension in order to test this hypothesis. In Piagetian terms, it would be argued that the genuine surprise of this discovery of temperature on the effect of enzymes on their substrates is sufficient to enable learners to produce 'schema' for 'reorganising' their understanding of this area of science, and in this way the information becomes 'internalised' – it becomes the learner's very own psychological property – and so learning is secured.

Even for the most able, it is difficult to see how this can be so. Solomon (1980) makes the point, for example, that in biological experiments controls are often required (as in the case of the enzyme experiments just referred to). But, in order to set up a control, children need to have a fairly good idea of what is going to happen. If they genuinely do not, and the teacher sets up the control, the control itself will act as a clue to the controlling factor, so that any element of surprise is pre-empted. Then again, the very fact that the children have been 'guided' towards the experiment will warn them that something of import is likely to happen. The trick of introducing some of the theoretical basics before the practical work is commenced ('well, they must know what they are supposed to be doing!') often gives the game away as well. This kind of practical work in effect often degenerates into no more than a following of a series of recipes, and such experiments would be more honestly entitled 'Experiments designed to show that...'.

In fact, the situation is even more removed from 'guided discovery' than this. As in the case of the maggot experiment described at the beginning of this chapter, salivary amylase and yeast do not behave as theory would have us believe. Both of these enzyme systems function most efficiently at temperatures higher than body temperature. This is such a 'surprise' result that it is usually ignored! The net result of all of this is that children learn to suspect, deny, refute or conveniently explain away unexpected results in a way that totally negates much of the purpose of this method of teaching science.

Whilst all this is true of most able children, it must also hold true for less able and the underachieving children. Perhaps it even contributes to the underachievement of those children who see through the charade and 'refuse to play the game'.

Practical work versus active learning methods

Whilst there is a long tradition of practical work in the science curriculum, especially since the 1960s' 'curriculum revolution', many teachers who advocate its extensive use are uncertain of its precise role. There is much confusion, too, in the educational literature regarding its real purpose. As a consequence, much practical work in schools is aimless, trivial and badly planned. It is used unthinkingly, without proper regard for its pedagogic or epistemological roles. Precise instructions are given, but the rationale underpinning the experimental procedure is often unclear, so that children acquire certain manipulative skills but little else. As far as less able children are concerned, what goes on in the school laboratory contributes little to their learning, and is of dubious value to their lives outside the classroom.

A major factor in this sorry state of affairs is that teachers attempt to serve several different purposes with a single experience. Generally, these purposes fall into four major categories:

- to stimulate interest and enjoyment
- to teach laboratory skills
- to teach the processes of science
- to assist the learning of scientific knowledge.

Since 1963, when Kerr carried out his now-classic study of practical work in secondary school science, there has been a significant shift away from a concern with learning scientific knowledge (rated first in 1963) towards the processes of science and the fostering of attitudes and interests (Gould, 1978; Beatty and Woolnough, 1982a,b). This shift in priorities, stemming largely from the influence of the Nuffield courses, also served to link practical work very strongly with discovery learning. As discussed above, 'discovery learning' is more often a bandwagon slogan for progressive educators than a carefully thought out pedagogic procedure. It is a notion that has done much to confuse a whole generation of science teachers about the precise purpose of laboratory activities. It goes without saying that a practical session that lacks clarity of purpose, or attempts to achieve several diverse goals simultaneously, may prove pedagogically useless.

A particularly common contemporary approach to practical work is summed up by the Nuffield phrase 'being a scientist for the day'. It is a slogan that has led many teachers to adopt a position whereby children are required to learn scientific knowledge by methods supposedly based on scientific methodology. But not all learning experiences need attempt to mimic scientific method. It is absurd to suggest that the quite separate aims of understanding the procedures or science and learning scientific knowledge require learners to be put in a situation where they have to learn the content through the method. There is a major conflict here between epistemological validity and the need for active learning methods (Hodson, 1985a). This conflict arises from a fundamental confusion in science curriculum design that assumes that the use of practical work in class is directly related to the experimental phase in the practice of science. It is significant that children also assume there to be a direct correspondence between school laboratory work and the activities of practising scientists (Hofstein *et al.*, 1980; Kyle, 1980).

We believe that teachers should make children aware that school experiments are conducted for a variety of purposes and do not necessarily represent a model of the scientific process. Learning about the nature of experimentation in science is but one of the goals of science education (and, paradoxically, is one that is not necessarily best approached via laboratory work). Experiments in science have three major aspects: purpose, procedure/technique, and results to be interpreted. There may be some advantages, in class, in considering each of these aspects separately – as 'pre-experiment', 'during-experiment' and 'post-experiment' goals – and in selecting learning methods at each stage that more closely match the issues under consideration. As far as the four major justifications of school practical work are concerned, it would seem that consideration of pre-experiment goals has value in bringing about an understanding of scientific methods, during-experiment goals have value in relation to motivation and manipulative skills, and post-experiment goals are suited to the learning of scientific knowledge.

Our views have much in common with those recently expressed by Woolnough and Allsop (1985). In strongly opposing the Nuffield-style discovery learning methods, they argue convincingly that teachers need to put theory back into practical work and, conversely, take practical work out of theory learning. A reworking of the relationship between theory and practical work leads them to a new rationale for laboratory work based on three fundamental aims:

- developing practical skills and techniques,
- engaging in problem-solving activities, and
- 'getting a feel for phenomena'.

If Woolnough and Allsop's aims are reinforced by a fourth one, the important step of

- exploring, elaborating and testing theoretical structures through experimentation.

we believe that they would constitute a suitable framework on which to base the practical aspects of the science curriculum. Allied to small group discussion work, this fourth approach would provide a powerful method for exploring the ideas of less able children, consolidating their understanding, judging alternatives and, in Kuhn's (1970) phrase, acquiring the currently accepted paradigm.

There is also a degree of confusion inherent in the assumption that practical work necessarily means laboratory bench work. Any learning method that requires the learner to be active, rather than passive, accords with the belief that children learn best by direct experience. In that sense, practical work need not always comprise laboratory experiments. If one reflects for a moment on the characteristics commonly associated with underachievers (table 2.1, p. 26), it is quickly apparent that laboratory work, which often makes demands on psychomotor skills, motivational levels and organisational abilities that these children do not necessarily possess, may at times be almost uniquely unsuited to them. It is with these children that other interpretations of practical work are most relevant. This is not an argument against providing laboratory experiences for less able children. It is an argument in favour of providing a range of alternative active learning methods.

Elsewhere, Hodson (1986c) has argued that a philosophically and pedagogically more valid science curriculum can be constructed by providing a 'mixed diet', comprising a variety of active learning experiences:

- practical 'exercises' for skills training
- teacher demonstrations and self-service video packages for the acquisition of specific content
- individual laboratory work for learning the activities of Kuhnian 'normal science'
- CAL materials for creative work involving hypothesis formation and testing
- open-ended inquiry for fostering enjoyment and interest, and for promoting an understanding of the practice of science
- project work for encouraging personal investigations and for providing the learner with opportunities for choosing the area of investigation, and for designing the investigation strategies and procedures.

As far as underachievers are concerned, we see particular value in the last three activities. All three make significant contributions to the learner-centred aims we identified in chapter 3 as having curriculum priority for these children. The learning advantages of open-ended inquiry and project work will be discussed in chapter 8 (pp. 180–2) and chapter 9 (pp. 196–9) and the associated organisational issues in chapter 8 (p. 166); those associated with CAL depend, of course, on the nature of the CAL materials employed.

Computer assisted learning (CAL)

Intelligent use of the microcomputer in science education depends rather more on the discriminating and critical application of educational criteria to the selection or design of software than it does on high-level programming skills. A discussion of how and how not to use computers in science education should be rooted firmly in psychological, philosophical, and curricular considerations. To guide and inform our practice we need a simple classification system and a simple theory of CAL based on such considerations. There are, of course, many ways of classifying and categorising, but for our purposes two are of major significance: the nature of the learning situation, and the psychology of learning underpinning the software design. By 'nature of the learning situation' we mean, literally, 'Who uses the computer?' There are two extreme categories:

- the teacher uses the computer, as a kind of 'electronic blackboard', to support and enrich other learning experiences;

- the individual child uses the computer as part of an individual-ised learning approach.

There are enormous advantages in the use of the computer as a sophisticated audio-visual aid (AVA) and we hope that teachers will take every opportunity to utilise and exploit the computer's potential in this respect. In contemplating this style of usage, teachers need to consider all the issues pertinent to the effective use of AVA and we recommend, in this context, the work of Kemp and Dayton (1985).

An equally important consideration is that of teaching style. Again, there are two extremes:

- the computer is used as an extension of the teacher (for performing quick calculations, manipulating data, plotting graphs, etc.) in teacher-directed activities;
- the computer is used as an alternative focus of attention –posing questions, checking answers, showing the consequences of making certain decisions, etc. – within a discussion-oriented teaching style.

Both approaches have value, and we would hope that both would be employed from time to time. However, we see the major value of CAL being associated with individual use, rather than teacher use. Because of our commitment to the belief that affective matters govern access to cognitive matters, we consider that giving children direct access to the computer, and allowing them to be in control, is a prerequisite to effective learning through the computer and to the attainment of computer literacy.

The classification system originally used by Kemmis *et al.* (1977) and elaborated by Rushby (1978) divides CAL into four 'paradigms': instructional, revelatory, conjectural, and emancipatory. The emancipatory use of the computer to reduce the workload of the learner or the teacher cuts across the other three categories. MacDonald (1977) distinguishes two kinds of labour in which the learner must engage: 'authentic' labour, which is an integral part of the learning and makes a valuable contribution to it, and 'inauthen-tic' labour, which is additional to the learning and of no particular value in itself. Because the computer excels at rapid, accurate calculation and information handling, it is an ideal tool for reducing the amount of inauthentic labour in the learning task. We would strongly advocate its use in such a role. The important educational decision focusses on the question of when, and in what circumst-ances, a task is classified as authentic or inauthentic. No generalisa-tions can be made, and each case must be considered on its merits. All we would ask is that the needs and interests of the learner are kept uppermost during the decision-making process and that the decision is based on the recognition that if learning is to be successful it should be as simple and straightforward as possible. Actually performing the calculation or searching through the data may encourage thought and may assist the establishment of ideas in the mind of the learner. But if such is not the case, then it is an inefficient use of the learner's time and, for many of the children we are concerned with here, it is likely to be counter-productive and to foster feelings of apathy or even hostility to the learning process. Another aspect of emancipatory CAL that is worth mentioning at this point is computer-managed learning. It may well be that the computer provides the only way in which the complex curriculum organisation we advocate in chapters 7 and 8 can be managed.

Each of the other three categories makes different assumptions about the learning process. In instructional CAL the computer presents the learner with material to be learned in small, easily assimilated chunks. The focus of attention is the subject matter and the learner's mastery of it. Revelatory CAL guides the learner through a process of learning by discovery, in which the subject matter and the underlying theory are progressively revealed. This category exploits many of the unique features of the computer as an aid to learning and enables the learner to be provided with learning experiences that are simulations of real life situations. In conjectural CAL the computer assists the learner to manipulate and test ideas and hypotheses. It is based on the assumption that knowledge can be created and re-created through the learner's experiences. Here the emphasis is on the learner's exploration of ideas, using the computer as support.

Instructional CAL is based firmly on a behaviourist psychology and assumes that precise learning outcomes can be specified in advance and that the knowledge to be acquired can be broken down into small units. The individual learning steps consist of three basic elements:

1. a stimulus provided by the program
2. a response provided by the learner
3. feedback from the program concerning the appropriateness of the response.

By reinforcing appropriate reponses and correcting inappropriate ones, the learner is led towards the desired learning outcome by a series of small steps. The appeal of this style of CAL is that learners can proceed at their own pace, can make their mistakes 'in private', can be provided with a 'tutor' of infinite patience, can be provided (via branching programs) with a learning route suited to their individual needs, and can have their successes reinforced and their misconceptions quickly identified and corrected. Its major disadvantages are that it is knowledge oriented and presents only a restricted amount of interaction. Used sparingly and appropriately, however, it has considerable potential for slow learners.

Unlike instructional CAL, where there is usually an alternative learning route, *revelatory CAL* frequently provides an opportunity for learning experiences that otherwise could not be provided (Sparkes, 1983). In the context of the science curriculum, this is most easily seen in its use in simulated experiments. Many experiments in science present complex problems, because they are too difficult to perform, too expensive or too dangerous, or because they operate on too long or too short a time scale. For example, learners can study the principles of genetic inheritance, the processes of mountain building and erosion, and the operating conditions of industrial plants quickly, cheaply and safely by means of simulations. Above all, they can do so in an environment that is under the teacher's control. In designing the simulation, the teacher can predetermine the level of complexity and the number of variables considered, thus tailoring the learning program to the specific needs, interests and capabilities of the learning group. Revelatory CAL, employed to encourage discussion within small learning groups, may foster a whole range of decision-making, communication, interpersonal and attitudinal skills (especially A1–A10 in table 3.2, p. 55) that are difficult to develop with other methods, but that are of paramount importance to our target group of children. By unfolding time at a quicker rate than 'normal' (except in the case of certain very rapid processes where the simulation actually 'slows them down'), by putting the learners 'in the hot seat' (where they have to 'live with' the consequences of their decisions), and by providing opportunities for collaboration and for learning from the successes and mistakes of others, this technique provides slower learners with a more stimulating and a more supportive learning environment than is usually possible. However, without adequate preparation and follow-up work, simulations can leave children with a sense of unreality, and in our view the major weakness of discovery learning is that through its emphasis on 'open-eyed and open-minded observation' (Hodson, 1986c), it ignores the learner's existing knowledge framework.

A particular strength of CAL is its emphasis on learner activity. In moving from instructional CAL to revelatory CAL, the capacity of the computer to involve the learner in a dynamic and responsive learning environment is exploited. In moving from revelatory CAL to conjectural CAL, the computer allows the learner to engage more frequently in the higher-order activities of problem solving and creative thinking. There are, of course, many similarities between the simulations in revelatory CAL and the modelling in conjectural CAL. Both involve using the computer to mimic some system, process, relationship or phenomenon so that its characteristics may be investigated and predicted, but there is one important difference. In a CAL simulation the learner is encouraged to change the external conditions but is prevented from altering the theoretical core of the model, whereas in conjectural CAL it is the theoretical model itself that is the subject of speculation.

The emphasis in *conjectural CAL* is on creative speculation and test. Thus, in a very direct way, it reflects the procedures described by Karl Popper as comprising scientific methods. Three examples should serve to illustrate this point. Programs such as ANIMAL assist learners to build up classification systems for themselves, by contemplating the characteristics that might be of significance to zoologists. Interrogation of chemical data bases, such as PERIODIC PROPERTIES, enables learners to hypothesise about chemical behaviour and quickly check out their predictions. At a more sophisticated level, the DMS program allows learners to construct and test theoretical models to account for real observations previously recorded. What is important in all these learning situations is that there is no 'hidden knowledge' in the computer. On the contrary, knowledge is located in the head of the learner and the computer is used by the learner to explore, modify and develop that knowledge. In other words, the starting point is the learner's existing knowledge, and the learning process is the reconstruction of that personal understanding in the light of feedback from the CAL package.

Viewed in this way, conjectural CAL fits perfectly into the constructivist psychology of learning and the emphasis on affective goals that we espoused in the early chapters of this book. Used as part of small-group learning, we see this form of CAL as being in many ways more productive than laboratory bench work for the children we are primarily concerned with in this book. Allowing the children to be 'in control' of the learning process is a major factor building the intellectual self-confidence necessary to raise the attainment of these children. Microprocessing for the less able child can be effective in restoring motivation 'because it offers features closely related to the learning needs of children. Failure [can be] made virtually impossible' (Johnson, 1986). Goldenberg *et al.* (1984) claim that for some children computers in themselves represent fascinating subject material. For others, they represent access to interesting activities, and, to still others, a change of routine.

None of this is to say that the 'state of the art' in CAL is as advanced as it might be. Schools are not using microcomputers in their teaching of science as frequently as it was predicted they would 20 years ago (Suppes, 1966). One of the major reasons for this is that education has not been able to take advantage of computer software's potential compared with traditional textbooks and laboratory procedures. There needs to be a shift away from a model of software writing that apes textbooks to take full advantage of such features as animation, personalisation of text, more appropriate use of graphics, and the highlighting of specially important concepts by the use of 'pause' statements, clock facilities, and flashing, for example.

Some very effective little programs have been written by our own students, most of whom have no more than a modicum of programming skills. Less able children can manipulate only one variable at a time (chapter 4). On this principle, programs do not need to be complex. One program, written for a 16K Spectrum, allows a child to change the temperature of the atmosphere by simply pressing the 'up' or the 'down' arrow on the keyboard. The temperature is recorded in the top right-hand corner of the screen. As the temperature increases, a lizard appears from under a stone, warmed by the morning sun, and climbs upon the stone to bask in the sun. If the child keeps on raising the temperature (which covers a range of 0–90° C), the lizard creeps back under the stone for protection from the midday sun. This simple device teaches that body temperature can be controlled by other than physiological means, i.e. by behavioural means.

Carrick (1977) has analysed a number of popular school biology textbooks in terms of the levels of organisation at which topics were approached. She describes six such levels: the molecular, cell, tissue and organ, organism, population, and community. In a later study (1982) she states that both American and British textbooks give most attention to the tissue and organ level. In a survey of biology software produced for the Acorn (BBC) microcomputer, Reid and Shields (1985) found that almost half the programs were concerned with population and community levels of organisation, and that only 6 out of 40 were concerned with the tissue and organ level. This is an interesting difference, for it implies that the two modes of presentation can complement each other, CAL giving valuable additional perspectives to the study of biology.

For a number of years now, the government has been encouraging the use of microcomputers in schools through its Microelectronics Education Programme (MEP), originally launched in 1980. This has now been succeeded by the Microelectronics Education Support Unit (MESU), which is planned to run for five years from 1986. Funded by the DES, it has an annual budget of some £3m. Part of its work is specifically allocated to the concerns of children with special needs, and to this end four Special Education Microelectronic

Resources Centres (SEMERCs) have been established, based in Bristol, Manchester, Newcastle-upon-Tyne and Redbridge. In addition, a Special Needs Software Centre has been established in Manchester. Teachers of the less able who are interested in the potential offered by this new technology are strongly advised to contact their nearest SEMERC, the addresses of which are given in the Appendix. Until 1989 at least, the SEMERCs will be developing software and distributing it freely to schools.

Here we will report on one of the most successful CAL devices called the concept keyboard. This device elminates the need for a QWERTY-type keyboard, and substitutes touch-sensitive pads. It is available in two sizes, to suit A3 (297 × 420mm) or A4 (210 × 297mm) paper. In essence it consists of a flat box about 1 inch thick, and is the size of either A3 or A4 paper. It connects to an Acorn (BBC) microcomputer via a ribbon cable, and currently costs £127 (A3 size) or £99.80 (A4 size); see figure 6.3a and 6.3b. A piece of ordinary A3 or A4 paper is placed on the pad by the teacher, who can then develop an overlay by using a piece of software called Starset (apply to your local SEMERC for details of availability). Alternatively, or in addition, a variety of overlays is obtainable free of charge from SEMERC. Figure 6.4 shows an overlay produced for teaching about the materials of which laboratory equipment is made. The child would first of all finger-touch the bottom left box 'THINGS WHICH', followed by the 'contain glass' box. These words will appear on the VDU. The child then touches all the pieces of equipment that contain glass, (e.g. test tube, thermometer, teat pipette), and a list appears under the appropriate heading. Eventually a series of three lists appears, which in turn can be printed out for sticking into a folder. It would be a simple matter to replace this with a picture of, say, a laboratory for the purpose of teaching about safety. Dyke (1986) gives a number of examples of how such overlays may be used to introduce children to real materials, such as circuit diagrams, plants picked on a field trip, sachets of chemicals or soils, or samples of rocks or machine parts. In the case of the plants, they are first pressed, and then stuck to the overlay. When a child touches, say, a root, information comes up on the VDU about the type of soil the root requires for optimal growth. Touching the fruit will cause information about its value as a food crop to appear (see figure 6.5). With the added facility of synthetic speech, this simple form of data base is ideal for use by visually handicapped children.

Unfortunately, these so-called 'framework programs' (i.e. 'strategy free' programs; King, 1986) are not widely used. They are quite different from what teachers are used to, and it is taking time for them to appreciate how valuable and flexible a teaching aid they can

Figure 6.3a Concept keyboard in use (Joe) with overlay, BBC microcomputer and VDU (photograph by David Griffiths, University of Manchester)

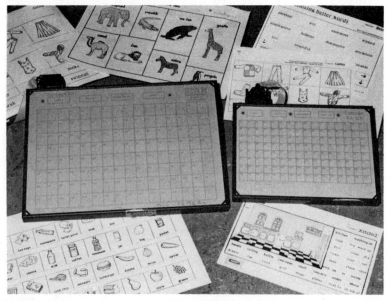

Figure 6.3b Keyboard and overlays (photograph by David Griffiths, University of Manchester)

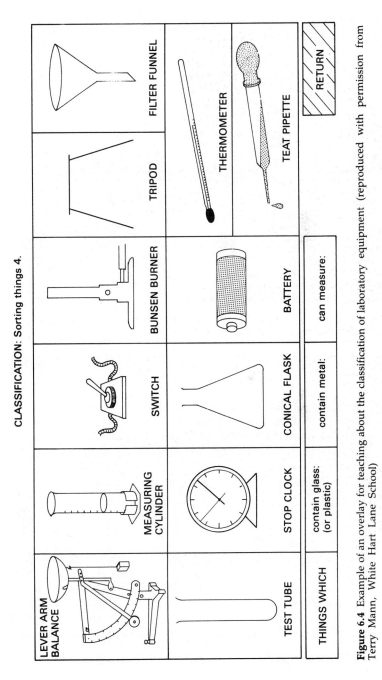

CLASSIFICATION: Sorting things 4.

THINGS WHICH	contain glass: (or plastic)	contain metal:	can measure:

LEVER ARM BALANCE

MEASURING CYLINDER

TEST TUBE

STOP CLOCK

SWITCH

CONICAL FLASK

BUNSEN BURNER

BATTERY

TRIPOD

THERMOMETER

FILTER FUNNEL

TEAT PIPETTE

RETURN

Figure 6.4 Example of an overlay for teaching about the classification of laboratory equipment (reproduced with permission from Terry Mann, White Hart Lane School)

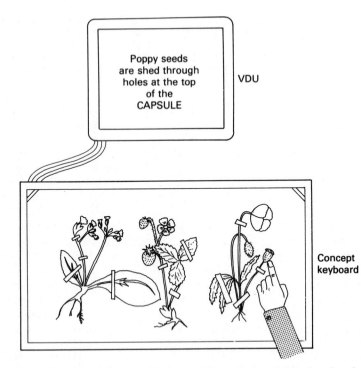

Figure 6.5 Example of the use of real materials on the concept keyboard – plants

be. With the release of free software and also the energetic way in which the SEMERCs, once approached, will offer advice in the use and development of such materials, they should become an indispensable aid to teachers of less able and handicapped children. Garrett and Dyke (in press) have produced an 'impact INSET pack', which includes extensive support materials for use with content-free software.

Methods of observation

One aspect of laboratory bench work that has, in the past, led teachers into a certain amount of confusion is the process of observation. Since the major reorientation of practical work during the 1960s, observation has often been given curriculum priority in science. This is well illustrated by the following extracts from the Teachers' Guides for Nuffield chemistry and physics.

Our essential aim ... is for pupils to enjoy their experimenting.

> For what they need are simple general instructions, where to look but not what to look for. (Revised Nuffield Physics, 1977)

> We want pupils to learn to distinguish between observed phenomena and explanations put forward by the creative thinking of the human mind. (Revised Nuffield Chemistry, 1975)

We want to question some of the assumptions underpinning this position, in particular the assumptions that observations provide a secure basis of fact, that scientific investigations commence with observation, and that observations can be independent of theory. In practice, observation is profoundly influenced by the observer's previous knowledge, experience and expectations. Even the conditions under which the observation is made may be significant. Thus, the colour of an object may depend on the nature of the illuminating light or the state of fatigue of the observer (different receptors are affected to a different extent by tiredness). This is not to say that there is no stability or permanence in observations. Their dependence on belief and experience is not such as to make observation totally unreliable and, as a consequence, science impossible. Terms in observation statements have their meaning located in the role they play in a theoretical structure. These theories (which give meaning to phenomena and events) necessarily precede observation. Without theories there are no concepts and without concepts no observations or precise observation statements can be made. It is not possible here to enter into a full discussion of the philosophical and psychological issues concerning the role and status of observation in science and interested readers are directed elsewhere (Johnson Abercrombie, 1960; Hodson, 1986a,c). Whilst there is considerable dispute between philosophers about the nature of observation, there is a measure of agreement that the three assumptions identified above as implicit in the Nuffield science courses are invalid. In other words,

- observations do not provide a secure basis of fact
- observations do not constitute the starting point for science
- observations cannot be theory independent.

Acceptance of this new position regarding observation has a number of implications for the curriculum. Rethinking and reappraising the role and status of observation in school science courses provides a number of urgent priorities (Hodson, 1986c):

- recognition that observation in science is unreliable and theory dependent
- realisation that the techniques of scientific observation have to be learned

- acceptance of children's existing conceptual frameworks as the starting point for science education
- reconsideration of the desirability of the discovery approach in the light of the dynamic relationship between observation and theory
- rejection of the objective and value-free image of science traditionally projected by the curriculum.

Optical, tactile and auditory illusions are particularly helpful in establishing the unreliability and theory dependence of observation, and film/video material using a moving camera establishes the important principle that the position and movement of the observer determines what we see. The important teaching point is that observation is a cognitive process and that in making observations we need to distinguish the relevant from the irrelevant. A start can be made through various kinds of observation games. For example,

'What is different about these two objects?'
'What is the same about these two objects?'
'Find an object which matches exactly the object on the desk.'

It is, of course, much easier to identify differences than similarities. Observation games are all very well and very useful, but they only become science when there is a purpose and a theoretical perspective. This transition from observation to scientific observation might be best achieved through various kinds of classification exercises. What is quickly apparent to children is that there is often a variety of criteria that can be employed, but that without criteria nothing can be classified. The skills acquired in classification exercises and in exercises concerned with the ordering of objects by size or mass, or by sequencing them in time, can lead towards the activity crucial to good scientific observation: the recognition of patterns and sequences. Again, it should become apparent to children that this requires a theoretical perspective. A range of activities involving the identification of objects by listening, smelling and tasting would help to reinforce the view that observation requires assumptions to be made and questions to be asked as a guide to the collection of data. Comparisons of actual behaviour with predictions based on those assumptions form the basis of the identification. Young's (1979) suggestions for the use of puzzle boxes would be particularly helpful at this stage.

The findings of the Assessment of Performance Unit (APU, 1985) indicate that children consistently under-use their senses: they are capable of fine discrimination and attention to detail, but do not make such observations unless their attention is carefully focussed. Nor do they readily identify significant relationships between observations without considerable guidance. In emphasising and promoting so-called 'open-eyed and open-minded' observation, much of the Nuffield-inspired contemporary science education misses the essential point about good scientific observation – that it depends crucially on education and training. Conducting scientific observations comprises a number of steps, each of which is theory dependent and each of which has to be taught. Learning experiences in school should be designed to give children awareness of and practice in each of these steps.

1. selecting the significant features and deciding what to look for
2. identifying, controlling and manipulating the variables
3. deciding on the equipment and materials needed
4. taking measurements, if appropriate and necessary
5. describing the observations
6. establishing links between individual observations and identifying trends and patterns
7. ensuring repeatability
8. achieving consensus through criticism.

If, as argued above, a particular theoretical background is a prerequisite for making valid and reliable observations, it follows that the child's existing conceptual framework is crucial. It also follows that children need to be given time and opportunity to reconstruct their understanding of scientific phenomena for themselves.

Even the selection of significant features can be a difficult task for the less able. Children taken to the zoo were asked to consider the role of the 'camouflaged' markings of the giraffes in an enclosure. This apparently simple concept seemed inaccessible to them, until it was realised that in the enclosure the markings had the reverse effect – the contrast between the light and brown areas of the fur actually made the animals *easier* to see. The fact that the lion (a predator) has only monochrome vision and sees only in shades of grey, and the breaking up of the outline of the giraffe, was better observed in diagrammatic form (see figure 6.6). Active observation is often encouraged by the use of diagrams (figure 6.7).

Hodson (1986a) has suggested that the complex relationship between theory and observation is best established, and the child's understanding is best developed, by the following series of activities:

1. making their own ideas explicit through writing and through discussion with other children and with the teacher

Figure 6.6 Lions observing giraffes (reproduced with permission from Chester Zoo)

2. exploring the implications of their ideas
3. matching and testing their ideas against experience – the teacher might well challenge them to find observational support for their ideas
4. using their theoretical ideas to explain observations
5. applying ideas to new situations
6. modifying and refining ideas to ensure better match with observations

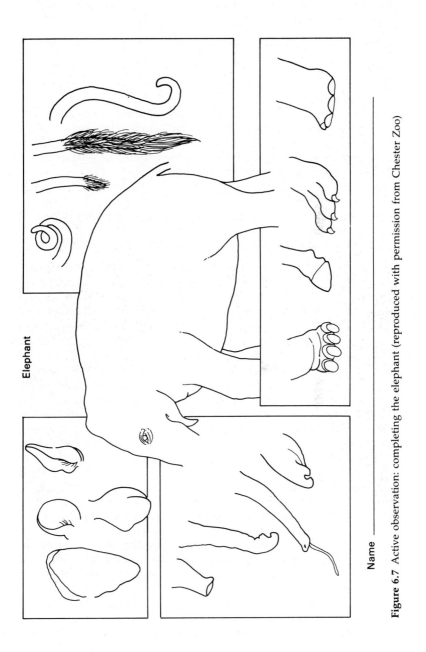

Elephant

Name

Figure 6.7 Active observation: completing the elephant (reproduced with permission from Chester Zoo)

7. making predictions and looking for observational support and test.

At this point the teacher would begin activities designed to effect a shift in understanding:

8. introduction of experiences to challenge and contradict children's existing views
9. encouraging the generation of alternative conceptual frameworks and explanations – even simple exercises such as 'Produce as many explanations as possible for why this bulb won't light' give children valuable experience of the brainstorming activities of hypothesis generation
10. introduction of the 'official' explanatory framework as one of these alternatives
11. exploration and testing of all alternatives (repeating steps 1–7)
12. comparison, judgement and choice, leading to consensus.

Steps 1–7 resemble the activities of Kuhnian 'normal science', while steps 8–10 resemble the events of Kuhnian 'revolutionary science' (Kuhn, 1970). At each stage, science teachers should be concerned with 'observation for a purpose'. That purpose may be simply the collection of data or it may be the illustration of an idea, the testing of a proposition, or whatever. Whatever the purpose, it should be made clear to the observer.

Such a detailed and carefully sequenced programme of activities is essential if children are to acquire the process skills (P1–P21) identified in chapter 3 (table 3.1). It is not possible to enter into a lengthy discussion of all 21 processes and comment is, therefore, restricted to one or two broad generalisations.

It is beyond dispute that children will not acquire these skills unless they are given experience of them and opportunities to develop them for themselves. Process skills such as identifying and clarifying problems (P1) and hypothesising (P2) require the provision of a rich variety of experiences and stimuli in order to provoke thought. But equally important is that children have the opportunity to discuss and to 'bounce ideas around'. Thus, small-group work would seem to be a prerequisite of process-oriented learning experiences. This has the added advantage of providing the security within which speculative hypothesising, highly imaginative experimental design, and daring interpretations of observations and data can be advanced. There is little chance of children being quite so daring in a class teaching environment. In all ot these creative processes of science – and we believe very strongly that so-called 'less able' children can and should be encouraged to be creative – the child is required to 'go beyond the known' and into the realm of 'what might be'. Only within a secure emotional climate will children take this step. The small learning group, engaging in CAL and other types of practical activities (as redefined above), is the place where this is most likely to occur. It is also the place where we are most likely to foster the attitudes (A1–A11) that we consider essential prerequisites if we are to raise levels of attainment for 'our children' (see table 3.2). The remaining attitudes (A12–A17) are more elusive and are likely to occur, if at all, from the cumulative experiences of the curriculum. We can do something rather more specific, however, through the more widespread use of project work, the inclusion of a degree of self-selected or negotiated content and the inclusion of contemporary social issues in the science curriculum, and, conversely, through the encouragement of scientific and technological considerations in the humanities curriculum.

7

Assessment and evaluation

DEFINITIONS

The curriculum model represented by the figure on p. 41 indicates the crucial role of the assessment and evaluation phase in obtaining and interpreting data on which decisions regarding curriculum modification can be based. The past two decades have seen such a considerable growth in the curriculum evaluation literature that Cooper (1976) is led to observe that 'evaluation has become one of the current catchwords of educational parlance'. However, despite this increased attention, there is still much confusion and, despite the vigorous encouragement of many LEAs, there is still a reluctance on the part of many teachers to engage in systematic curriculum evaluation. This reluctance stems, in part, from the confused nature of many of the LEA guidelines (Clift, 1982; Lusty, 1983) and, in part, from teachers' tendency to concentrate their efforts on the curriculum development phases in which they feel most secure: namely, the selection of content, the design of learning experiences, and, often retrospectively, the articulation of curriculum goals. Empirical support for the assertion that teachers afford a low priority to systematic evaluation has been provided by Hodson (1986b), using a questionnaire survey of 105 secondary school science departments. What is clear from this study is that the major impetus for curriculum change is the head of department's hunch that change is needed, augmented each summer by external examination results. Whilst acknowledging the value of experienced teachers' intuitive views, we believe that curriculum decisions informed only by such views are liable to be suspect. Just as suspect are decisions taken on the basis of 'bandwagon jumping', pressure from articulate pressure groups, or the glossy advertising copy of publishers. If teachers have invested considerable time, energy and thought in the design and execution of curriculum experiences, they should not devalue that investment by engaging in hasty and ill-designed evaluation.

A first step towards the encouragement of more systematic assessment and evaluation in school science departments is increased clarity regarding their nature and purpose. Some of the present confusion arises from the tendency of many teachers to use the terms assessment and evaluation synonymously. As far as this chapter is concerned, assessment refers to the process of obtaining information about a pupil's developing knowledge, abilities, skills and attitudes, while evaluation is concerned with the collection and interpretation of information on the basis of which decisions about the worthwhileness and effectiveness of the curriculum can be made. One criterion of 'effectiveness' is, of course, the extent of learning brought about in the pupils, so that assessment is part of the evaluation procedure.

It is not intended, here, to enter into a prolonged discussion of the various models of curriculum evaluation – interested readers are referred to the excellent reviews provided by Tawney (1976) and Skilbeck (1984) – save to note that the various processes of curriculum evaluation may, for convenience, be grouped into three major categories:

1. Appraisal of teaching and evaluation of the quality of classroom experiences.
2. Evaluation of curriculum goals, materials, learning experiences, and assessment procedures.
3. Assessment of learning gains and diagnosis of specific learning difficulties.

It is also worth pointing out that information gathered during these evaluation activities has two major functions: to guide the further development of the curriculum (formative evaluation) and to appraise the finished programme (summative evaluation). These terms, deriving from the work of Michael Scriven (1967), may also be applied to assessment. Thus, formative assessment provides detailed feedback to learners, teachers and parents, and seeks to recommend appropriate curricular action, whereas summative assessment provides a final record of a pupil's attainment and may be of value to potential employers, further education institutions, etc. Whilst these distinctions are useful, there is considerable overlap. Thus, summative evaluation of a 'curriculum package' in use in school A may provide information of value during school B's development activity, and so have formative value. Indeed, this is the kind of co-operative evaluation that we would wish to see much more in our schools. An end-of-year summative assessment may have the effect, indeed it may have the express purpose, of providing a stimulus to the pupil in relation to the coming year's work, and so have formative value.

APPRAISAL OF TEACHERS AND TEACHING

The first of the categories of evaluation activities listed above has as its focus of attention issues such as the quality of teacher–pupil relationships, the extent of pupil participation in lessons, the language of instruction, class management and the expertise of the teacher in various teaching techniques. These kinds of issues cannot be judged simply by consideration of the curriculum plan and the learning materials. They are best studied by classroom observation techniques, aided by the kind of observation schedules frequently used in initial teacher training and employed in some schools for the appraisal of student teachers and probationers (see table 5.1).

It is our view that teacher appraisal, concentrating on the kinds of issues listed here, is an essential part of systematic and comprehensive evaluation, and that science departments should consider, as a matter of some urgency, the adoption of regular appraisal. The video recording facilities now commonly available in many schools enable teachers to use such methods for self-evaluation, without the potentially threatening presence of an observer. Oldroyd, Smith and Lee (1984) have made some helpful suggestions about the organisation of such procedures and Walker and Adelman (1975) provide some useful theoretical perspectives.

Figure 7.1 contains a checklist of important issues and points to look for when observing teachers at work. Observation methods may profitably be supplemented by questionnaires that focus on matters such as children's enjoyment and staff satisfaction, together with observation of child behaviour. Clearly, much can be ascertained about the quality of the curriculum and the quality of teaching by observing children. We would anticipate that children following the kind of curriculum advocated in this book would show a much reduced incidence of the kinds of behaviours commonly associated with underachievers that we identified in chapter 2.

An alternative style of classroom observation directs attention not on teachers' expertise in the technical aspects of teaching but on their implicit psychological and philosophical stances. Some years ago, Eggleston, Galton and Jones (1976) claimed to have identified three styles of science teachers: those who are 'actor managers'; those who emphasise fact finding and fact gathering (the most common); and those who emphasise pupil-centred activity (the least common). To an extent, teachers in the study were able to change their style to suit the topic, but this change was never sufficiently marked to warrant recategorisation. More recently, Dibbs (1982) classified science teachers into four distinct types, based on their implicit philosophical positions: inductivist (I);

hypothetico-deductivist (H); verificationist (V); and those with no discernible beliefs about the nature of science. He claims that when parallel groups of children were experimentally provided with I-, V- and H-style teaching by Dibbs himself there were significant differences in children's understanding of the nature of science and scientific inquiry. He also found some evidence that V-style teaching resulted in higher scores on attainment tests but in lower recruitment to optional courses. Perhaps this serves to illustrate what many teachers know already: that successful teachers are those who are capable of adapting their style and approach to suit the situation and purpose. As far as our target group of children is concerned, this means – if we are correct in our earlier assertions – that teachers need to be able to switch their approach from the kinds of activities commonly employed with high-ability learners to the activities discussed in chapter 5 as particularly appropriate to the less able (see p. 88 on the use of DARTs activities for one example).

Dibbs used interviews with teachers as his major means of categorisation, whilst Eggleston *et al.* used a very elaborate schedule derived from Flanders' Interaction Analysis system. A more accessible set of guidelines for classroom observation and, more particularly, for self-evaluation by busy teachers has been produced by Roger Hacker. His *Science Lesson Analysis System* (Hacker, 1982) focuses on 12 categories of intellectual activity in science, ranging from simple recall and making of observations, to hypothesis generation and experimental design. Figure 7.2 summarises the major categories. Hacker's system, together with the guidelines in figure 7.1, should provide a sufficient basis for evaluating the quality of classroom activities.

EVALUATION OF CURRICULUM MATERIALS

Whilst the major sources of information for curriculum evaluation are peformance by pupils on end-of-topic tests of various kinds, teacher observation of classroom activities, interviews and questionnaire returns, there is much valuable information that can be gathered by critical scrutiny of curriculum plans by experienced teachers following systematic procedures. The major items for attention are the aims and objectives of the curriculum unit, the content, the learning methods (in particular, their appropriateness to the goals, content and children), the assessment scheme and certain general issues relating to cost, staffing, pupil and teacher satisfaction, etc.

Clearly, the precise information required by a teacher depends on the situation: the size, sex, composition, age and ability of the learner groups; the particular characteristics of the school, its laboratories

Name:	School:
Class/Ability group:	Date:
Subject/topic:	

	FOCUS OF ATTENTION	POINTS TO LOOK FOR	OBSERVER'S ASSESSMENT/ COMMENTS
PLANNING	1. Learning goals/ experiential goals	Were the aims and objectives clearly identified and appropriate?	
	2. Organisation and structure of lesson	Was the lesson well designed from the point of view of matching aims/content/methods/ assessment techniques?	
	3. Selection of content	Was the content appropriate to the pupils? Was it accurate, relevant, up-to-date and interesting? Did the content take account of different needs, interests and abilities?	
	4. Selection of learning methods	Did they take account of individual differences? Were they appropriate to the content? Was there sufficient variety of learning methods? Were the methods well sequenced?	
	5. Preparation of materials	Worksheets, study guides, etc.	
	6. Laboratory organisation and procedures	Were they appropriate to the class? Were they appropriate to the content? Were they appropriate to the lab? Were they safe?	
PERFORMANCE	7. Start of lesson/introduction	Impact of initial activity. Was the class quickly settled? Was the introduction too long/too short, interesting/tedious, etc.? Was the purpose of the lesson made clear to the children? How good was the judgement about children's background knowledge, abilities, etc.? Did the children know what to do, how to do it, what standards were expected?	
	8. Development of lesson	Did the lesson develop logically and purposefully? Was the teacher able to modify in the light of feedback? Were the key points isolated and emphasised?	
	9. End of lesson	Was there an adequate summary? Were conclusions drawn? Were the goals achieved? Did the children realise the point of the lesson?	
	10. Use of (i) materials/ equipment (ii) blackboard (iii) other AVA	Materials and equipment well organised and appropriate? Was the layout well planned? Was the writing legible? Was it free from spelling errors? Were they effectively and appropriately used?	

Figure 7.1 Important points to look for when observing teachers at work

PERFORMANCE (continued)	11. Competence with subject matter	This may range from inadequate to imaginative and inventive!	
	12. Pupil involvement and activity	Were the children actively involved? Did the lesson bring about significant learning or did it merely 'occupy the time'? Was suitable provision made for those who finished quickly, couldn't cope, etc.?	
	13. Pace of lesson	Too fast, too slow, about right? Were there changes in pace and tempo?	
	14. Timing	Relates to 9. Was the lab left in a reasonable state for colleagues?	
MANAGEMENT OF CHILDREN	15. Discipline	Was the teacher able to establish and maintain conditions suitable for learning with a minimum of fuss? Was the teacher able to secure and maintain attention and interest? Were there signs of boredom? Was there excessive noise? Was the teacher effective in dealing promptly, quietly, fairly and firmly with misconduct? Was the teacher 'in charge'? Was the teacher effective in anticipating and avoiding behavioural problems?	
	16. Communication	Clarity, pitch, inflexion and audibility of voice? Pace of delivery? Appropriateness of vocabulary? Enthusiasm and liveliness of manner? Are pupil–teacher relationships sound? Does the teacher listen to the children?	
	17. Catering for individuals	Did all groups/individuals get a fair share of attention, guidance and encouragement?	
	18. General comments on style and 'presence'	Is the teacher purposeful and organised? Is the teacher relaxed, tense, nervous? Is a sense of humour evident? Is a sense of compassion and tolerance evident? Is the teacher generally supportive of the children? Any tendency to overuse/neglect a particular style or approach?	
EVALUATIVE TECHNIQUES	19. Questioning techniques	Were the questions apt? Were the questions well phrased? Were they distributed well? Were correct, incorrect and partial answers dealt with constructively?	
	20. Use of criticism and praise	Was praise and encouragement given for individual or group achievement or effort? Was constructive criticism offered?	
	21. Diagnosis	Was the teacher able to identify those experiencing difficulties? Was appropriate remedial action taken?	
	22. Homework	Was it appropriate, realistic in demand and adequately explained?	
	23. Marking	Is work regularly and constructively marked?	

Figure 7.1 (*continued*)

24. Any significant incidents?	
25. General comments, specific strengths & weaknesses, etc.	

Observer's assessment. If it is helpful, assessment may be coded:
N/A	Not applicable to this lesson
1	Weak. In need of attention
2	Adequate
3	Good
4	Outstanding

Figure 7.1 (*continued*)

and staff; and so on. For example, there are a number of schools in our large urban areas in which the IQ distribution is skewed markedly to the left with a mean value of approximately 90. When this is compounded by acute social deprivation and very high youth unemployment, the school clearly has massive problems of low pupil aspirations and, consequently, low attainment. In these cases priority must be given to the task of raising aspirations through the enhancement of pupil self-image and the generation of a climate of success, as discussed earlier. It is vital, therefore, that these schools recruit teachers who appreciate the special needs of the school and are sensitive to the fundamental needs of the children. Without the 'right kind of teachers', all plans for raising the levels of attainment of less able and socially disadvantaged children are doomed to failure.

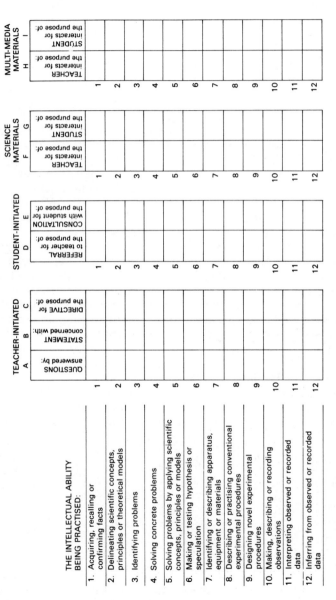

Figure 7.2 The science lesson analysis system (reproduced with permission from Hacker, 1982)

Despite the special problems of some schools, there are, nevertheless, certain questions of universal applicability. The generalised checklist in table 7.1 was designed by Hodson (1986d) as an aid to teachers presented almost daily with evaluation demands. He points out that

> some questions are at a more fundamental level than others, and a 'no' at a certain point renders the completion of the checklist unnecessary. Thus, the checklist is not simply a device for naively establishing the numerical ratio of 'yes':'no' responses. Rather, it provides a structure and an aide-memoire for the evaluation process by focussing the teacher's attention on key issues rather more sharply than might otherwise be the case.

We anticipate such a checklist being incorporated into an overall strategy for departmental review and management, utilising materials such as those recently produced by the ASE (Hull and Adams, 1981, 1982; ASE, 1985). Use of the checklist might assist teachers to identify specific weaknesses and deficiences in the existing curriculum and to formulate guidelines for the design and development of replacement curriculum units. Many curriculum materials fail to take account of the kinds of issues raised in chapters 2 and 4 as central to learning experiences for the less able. This checklist should assist teachers to focus quickly and efficiently on issues of language, activity, sequencing, matching, and so on. We believe that teachers already engage in such evaluation, but they do so intuitively and rather unsystematically. The checklist encourages a more systematic approach and thereby minimises the risk of overlooking key issues for the less able learner.

Replacement units for those identified as deficient or unsatisfactory might be produced in school, or they might be obtained from outside – from other schools, from publishers, from teachers' centres, or the like. We would want to encourage far more inter-school co-operation on the production of curriculum materials. There is too strong and long-standing a tradition of isolation in our education system. There is nothing to be lost, and everything to be gained, from close co-operation between teachers, in all aspects of curriculum design.

Hodson (1986d) envisages a second evaluation stage, the evaluation of the proposed replacement unit, providing the additional information required for the 'judgement stage' (figure 7.3), which includes the following steps:

1. Judge the additional demands in terms of facilities, equipment, staffing, time, cost (both implementation and running cost).

2. Judge the probable advantages over the existing unit.
3. Judge the probable disadvantages compared with the existing unit.
4. Compare with other alternatives.
5. Judge the feasibility of implementing the chosen unit.

Once a decision in favour of a particular option has been made, attention switches to strategies of innovation, implementation, and institutionalisation, taking into account the special 'school factors' referred to earlier. A discussion of these strategies is beyond the scope of this book and interested readers are directed elsewhere (ASE, 1985; Hull and Adams, 1981, 1982).

ASSESSMENT OF ATTAINMENT AND DIAGNOSIS OF LEARNING DIFFICULTIES

Deale (1975) suggests that, when evaluating an assessment scheme, teachers should ask five key questions:

• will making this assessment benefit the education of the children, directly or indirectly?
• is it a valid test of what they have been learning?
• can it be marked fairly or uniformly?
• will it provide, when needed, all or part of the appropriate information about the children's attainments?
• are there any important aspects of the course which are not covered by this assessment?

It has to be admitted that much of the assessment currently practised in secondary school science departments does not stand up well to this kind of scrutiny. Many science departments engage in haphazard and confused assessment procedures, which yield information of little or no value. By failing to provide adequate feedback for diagnosis or for the modification of learning experiences, these procedures often fail to facilitate learning. Many aspects of science courses, particularly non-cognitive ones, are totally ignored. Therefore, assessment schemes cannot provide a balanced picture of each child's strengths and weaknesses. The following is an attempt to summarise some of the inadequacies and harmful side-effects of current, traditional assessment methods (Hodson, 1986b).

• the use of test-bases stereotypes by teachers, and the pupils' acceptance of that stereotyping, leading to self-fulfilling prophesies and the further development of an adverse self-image for large numbers of children, particularly those who are our concern in this book

Table 7.1 *Checklist of questions for the evaluation of curriculum materials (Hodson, 1986d)*

I	*Aims and Objectives*
I.1	Are they clearly stated?
I.2	If not explicitly stated, are they implied in the nature of the materials?
I.3	Are they compatible with the existing curriculum?
I.4	Are they educationally worthwhile and desirable?

II	*Content*
II.1	What concepts, skills and attitudes are covered?
II.2	What concepts, skills and attitudes are assumed as prior knowledge?
II.3	Will the pupils have that knowledge?
II.4	Is the subject matter accurate and up to date?
II.5	How well does it relate to previous and forthcoming curriculum units?
II.6	How well would it meet the demands of external examination syllabuses (if relevant)?
II.7	Is the unit well designed from the point of view of overall structure, sequencing and level of cognitive demand?
II.8	Is the content suitable, in terms of difficulty and interest, for the particular age group under consideration?
II.9	Is it suitable for the ability range under consideration?
II.10	Does the unit attempt to relate the topic to other topics and to other subjects, particularly to social and economic issues?
II.11	Are the underlying assumptions about the nature of science and its methodology acceptable?
II.12	Does the unit provide satisfactory experiences of the processes of science (observing, measuring, recording, hypothesising, etc.)?
II.13	Do the materials reflect a particular or restricted cultural, social and political viewpoint?
II.14	Is there any tendency towards prejudice and stereotyping?
II.15	Are the materials well designed from the point of view of appearance and readability?
II.16	Do the materials exploit a suitable range of media?
II.17	Is there support material available or could it be readily produced?

III	*Learning Experience*
III.1	What are the main learning methods advocated?
III.2	How suitable are they for the content?
III.3	How suitable are they for the pupils under consideration?
III.4	Is there scope for pupils to work at different rates?
III.5	Are there appropriate remedial/enrichment activities?
III.6	Is there scope for pupils to work individually/in pairs/in larger groups?
III.7	Is discussion work between pupils encouraged or required?
III.8	Is there a variety of learning methods?
III.9	Are certain learning methods over-used?
III.10	Are pupils likely to be motivated by the methods?
III.11	Is there recognition that pupils need to be active in pursuit of their learning?
III.12	Is there a satisfactory role for practical work?
III.13	Are the experiments well chosen (manipulative skills required, considerations of safety, time and cost, etc.)?
III.14	Is a suitable range of thinking skills covered?
III.15	Do the writing tasks include a range of purposes and audiences?

Table 7.1 (*continued*)

III.16	Are the pupil tasks suitable for the pupils under consideration?
III.17	Does the unit provide for the pupils to accumulate a written record of the work done?
III.18	Do the materials and experiences contribute to the linguistic development, mathematical development and personal development of the pupils as well as their scientific development?
III.19	What is the role of the teacher? Is there any change from existing practice?
III.20	Is there any scope for children to exercise choice about the route and the activities followed?
III.21	Is there a satisfactory balance between factual knowledge, methods of processing and manipulating concepts, application of ideas to new situations, etc.?

IV	*Assessment*
IV.1	Is there a well-designed assessment scheme?
IV.2	Is there compatibility/conflict between the curriculum goals, the learning experiences and the assessment scheme?
IV.3	Does the unit include an effective scheme for assessing the achievement of objectives?
IV.4	Which of the learning objectives are assessed, and how?
IV.5	Is the whole range of thinking skills (as defined in III.14) tested?
IV.6	Does the unit include an effective scheme for diagnosing learning difficulties?
IV.7	Are there suggested criteria for the marking of pupil assignments?
IV.8	Are the tests fair? Do they give a fair measure of the success of the unit? Are the tests valid and reliable?
IV.9	Are the tests easy to mark?
IV.10	Are the 'test situations' well chosen?
IV.11	Is underachievement identified and appropriate remedial action taken?

V	*General*
V.1	Are any existing evaluation data available?
V.2	Does the unit allow teachers and children a degree of flexibility in approach?
V.3	Could it be adapted to the school's particular needs and conditions?
V.4	Are the materials durable and robust?
V.5	How difficult would it be to teach?
V.6	What additional resources and equipment would be required?
V.7	Would additional staffing or accommodation be required?
V.8	Would special timetabling be necessary?
V.9	Are there any retraining implications?
V.10	Could it be implemented immediately?
V.11	Is it possible to have a trial run of part of the unit and, if necessary, to abandon it?
V.12	What are the likely reactions of other departments, parents, employers?
V.13	What is the likely cost (implementation and running cost)?
V.14	What are the major strengths and weaknesses?

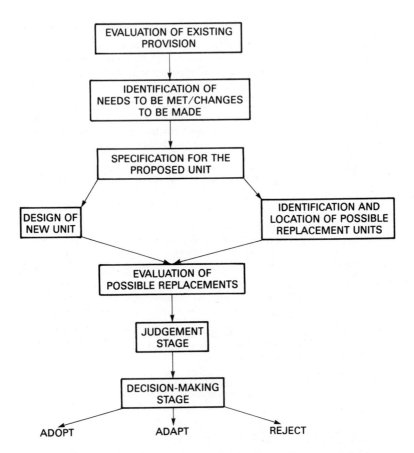

Figure 7.3 The processes of curriculum evaluation (materials)

- an undesirable emphasis on the assignment of grades and rankings, which fails to recognise and reinforce the achievements of the less able
- the encouragement of competitiveness at the expense of co-operation
- an over-reliance on norm-referenced testing, which identifies large numbers of children as failures
- a failure to provide children with adequate information about themselves and for themselves; assessment is seen – by the children – to benefit the school administration rather than the child
- a bureaucratic and impersonal climate for assessment, which attempts to achieve 'fairness' by uniformity, objectivity and

anonymity, and which makes little provision for individuality, creativity and interpersonal skills

- a concentration on academic issues, with a consequent failure to value non-cognitive aspects of the curriculum; in schools, it is often the case that non-evaluation implies low status in the eyes of both pupils and staff
- over-reliance on particular, limited assessment methods (usually multiple-choice tests and essays), which devalues other skills that children may possess and leads to examination-oriented teaching
- the inadequacy of recording and reporting systems that conflate detailed information into almost meaningless grades and scores, and fail to alert teachers, parents and children to learning problems.

If comprehensive schools are genuinely concerned for the well-being of all children and with their all-round development, as they usually claim, then that concern should be manifest in the assessment procedures. Good curriculum planning and good teaching demand good assessment procedures. In the introduction to part II we discussed the principles of systematic curriculum planning, the processes of which were represented diagrammatically. Adaptation of that model to the particular issue of assessment identifies a number of key questions.

- Why do we assess?
- What should we assess?
- How should we assess?
- How should we interpret test data?
- How should we respond?

Why assess?

The 'why' question has a number of answers. We seek to measure individual attainment in order that we can monitor progress, diagnose learning difficulties and identify learning strengths, and we need information on group attainment in order to evaluate the effectiveness of particular learning episodes. It is only through diagnosis that we can hope to raise levels of attainment. As discussed in chapter 2, there are many factors contributing to low attainment – problems of motivation, reading and numeracy skills, concentration, behaviour, etc. Too often, in the past, teachers have employed gross explanations of low attainment in terms of low IQ or insufficient effort. A more careful, individualised approach to the identification of learning failures is a *sine qua non* of an individualised curriculum.

Assessment data may be used to inform a variety of decisions. Amongst them are decisions about the grouping of children, the deployment of teachers, and the modification of learning experiences. Assessment data may be used as the basis of counselling advice to children and parents regarding subject choice or career, and to assist the compilation of reports and references. It goes without saying that lower-attaining children, and those with learning and behavioural problems, need more counselling and the assembly of more detailed data for diagnosis.

There is a school of thought that suggests that regular assessment acts as a motivator for both pupils and teachers. Indeed, the use of assessment (homeworks, assignments, and tests) to encourage learning is a long-standing tradition in schools. The belief that teachers are also, to a large extent, controlled and motivated by the external examination system is not without some foundation (indeed, teachers frequently complain that the examination system requires them to concentrate on knowledge-oriented activities at the expense of more open-ended activities), and we hope to show later in the book how this tendency can be exploited for the benefit of slow-learning children. As far as pupil motivation is concerned, we are not convinced that the competitive aspects of assessment are beneficial to any but a very small minority of high-ability children; we certainly do not see any benefit accruing to the children who are the particular concern of this book. The almost exclusive encouragement of competition, and the over-reliance on norm-referenced testing that is common in most schools, identifies large numbers of children as failures because it fails to recognise and reinforce the achievements of the less able. As a consequence, many children have their already poor self-image further damaged.

The traditional emphasis on grading has tended to concentrate the attention of teachers on techniques for discriminating between children, rather than on identifying learning gains and areas of difficulty experienced by individuals. However, if assessment is seen as the process of providing individual learners with meaningful information about themselves, and for themselves, we believe that it can act as a motivator. What is needed is a change of orientation for school assessment programmes. We should seek regular, systematic feedback to provide children with information enabling them to identify their strengths and weaknesses, and to furnish advice on how best to proceed. Similarly, we should seek to provide parents and other teachers with such information. The idea of learning and behavioural contracts, resulting from a weekly negotiation between child and tutor, has considerable value for these pupils in providing short-term, achievable targets. Once children feel emotionally secure and experience success in their science learning, there is likely to be a 'transfer effect' to other curriculum areas. One of the more obvious 'transfer effects' is on school attendance. In this context, a Manchester school reports an increase in average attendance from 74 per cent to 92 per cent following the introduction of short-term goal-oriented teaching, supported by profiling and accreditation schemes. We shall return to these matters in chapter 9.

What should we assess?

Clearly, we need to know in advance what we intend to assess, not only in advance of the assessment process itself, but also in advance of the teaching. It would be a cardinal error to wait until the teaching/learning process was complete, and then seek to ascertain the outcomes. We need to have expectations about acceptable levels of attainment in certain specified areas. Indeed, the expected and desired outcomes to a large extent determine the content and the methods of the curriculum. In the same way, the learning goals themselves should constitute the framework for the construction of the assessment instruments. There are, however, two major traps for the assessment scheme designer to avoid. First, as pointed out in chapter 3, we must recognise the difference between the learning goal and the evidence that we seek as an indication that the goal has been attained. And, second, we must resist the temptation to concentrate so much on the extent to which children attain certain pre-specified goals that we fail to recognise the unexpected, the idiosyncratic and the creative. In other words, we need to design an assessment strategy in which the qualities we value, and which we seek to produce in children, will be recognised and, at the same time, we need to build into it the capacity to respond to the unexpected.

Traditionally, assessment procedures have concentrated on cognitive matters, whilst non-cognitive aspects of the curriculum have often been ignored. Consequently, a balanced picture of each child's particular strengths, weaknesses and dispositions has rarely been obtained. In the drive for 'fairness' through uniformity, objectivity and anonymity, a bureaucratic and impersonal climate of assessment has been created that makes little provision for individuality, creativity and interpersonal skills. This state of affairs must not be allowed to persist. It is precisely the personal element of the curriculum that we have argued is the key to raising the level of attainment for less able children, and that should therefore have priority in the assessment scheme.

In a well-designed scheme, the assessment procedures would be compatible with the principles employed in planning and implementing the curriculum. In other words, there should be harmony

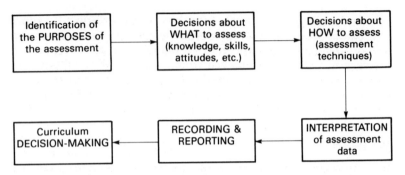

Figure 7.4 A model of the assessment process

between the educational goals, the learning experiences provided, and the assessment and evaluation procedures. We should assess what we intended children to learn, using methods appropriate to the curriculum philosophy, and we should ensure that such is the case by formulating a test specification based on the learning objectives and by employing assessment based on the same kinds of activities as the learning methods. This 'rational' approach to the assessment procedure, represented diagrammatically in figure 7.4, proceeds from a consideration of the purposes of assessment, through a discussion of what and how to assess, to a concern with interpretation, recording and reporting. However, there is evidence (Hodson, 1986e) that many teachers start by writing the assessment items, then proceed to classify them and, finally, identify the assessment objectives. What they rarely, if ever, do is to consider the purpose of the enterprise. This is strikingly similar to the tendency of teachers to decide what to teach (the content) and then, retrospectively, to identify their curriculum goals (Reid and Tracey, 1985).

How should we assess?

As far as the curriculum proposals in this book are concerned, there are three broad areas to be included in the assessment scheme: knowledge, processes and attitudes. Clearly, each of these aspects of the science curriculum is best assessed by specialised techniques.

Immediately we begin to ask how to assess, we are beset with alternatives: formal or informal assessment, terminal or intermittent, course work or examinations, and so on. Thus, the question of how to assess becomes intimately bound up with questions about the design and organisation of the assessment programme. There are no easy and certain answers to these questions. However, two valuable sources of advice on how to set up a thorough and detailed assessment scheme are Sutton (1986) and Ward (1980). What we can expect from a well-designed scheme are relevance and appropriateness (validity), and reliability. As far as validity is concerned, the golden rule is that the assessment method must suit the purpose and content of the assessment. For example, multiple-choice tests are hardly relevant to the assessment of a child's social interaction skills! A driving test that seeks to establish whether drivers can identify and respond appropriately to certain traffic signs is more valid than one that asks whether they can describe such signs. A test is valid if it uses an appropriate technique and adequately samples the content of the course, and so measures what one seeks to measure. It is reliable if the same test performance produces a similar score on different occasions, or with different markers.

	Terminal	Intermittent or periodic	Continuous
Knowledge	written tests	written tests, projects, assignments	coursework, discussions
Processes	written tests practical tests	written tests, assessed practicals, projects, assignments	coursework, observation
Attitudes	questionnaires	projects, observation, checklists	observation, discussion and interview

Figure 7.5 Main assessment methods

Regarding the curriculum proposals we have advanced in this book, we would recommend both terminal and intermittent assessment, because we have both summative and formative assessment needs, and we would advocate the use of a variety of assessment methods and instruments (figure 7.5). Reseach shows that higher levels of attainment are obtained when courses employ frequent testing, so we would advocate regular, end-of-topic tests, together with 'assessment situations' based on normal lesson activities (see later) and informal assessment using observation and discussion methods. In a book of this kind it is not possible to provide detailed considerations of assessment methods. The best we can do is to formulate guiding principles in each of the three main areas. Detailed technical advice can be obtained from the several excellent textbooks dealing with assessment issues written specifically for teachers constructing their own tests (Satterly, 1981; Harlen, 1983; Frith and Macintosh, 1984).

Knowledge

Because we have not identified the specific content of the curriculum, for reasons outlined in chapter 6, we are not in a position to identify the specific content of the assessment scheme. However, we do feel able to put forward certain general guidelines to act as a framework.

	SYLLABUS CONTENT (by module)			
	1.1	1.2	1.3	1.4 etc.
Concepts & terminology				
Facts				
Trends & sequences				
Classifications				
Principles & generalisations				
Theories				
Experimental techniques				
Societal & ethical issues				

Figure 7.6 Planning matrix for test design (first step)

The first step in designing an efficient assessment scheme is the identification of the specific knowledge content, using a guide such as that in figure 7.6. The matrix serves to remind teachers that in planning tests it is important to sample content systematically and comprehensively. It is envisaged that this same planning matrix is employed in the process of designing the learning experiences.

A major principle underpinning a good assessment scheme is that it does not concentrate solely on scientific knowledge. It should test children's awareness of other ideas and principles implicit in the curriculum design – ideas derived from the philosophical and psychological stance of the course, from the emphasis on contemporary social, economic and environmental issues, and so on. It is possible to identify one or two general issues of these types, but their precise nature will, of course, vary from school to school. Examples of the kinds of things we have in mind are:

- observation is unreliable and theory dependent
- techniques of observation have to be learned
- theories may be realist or instrumentalist
- scientific knowledge is tentative
- hypotheses are tested by experiment
- science and technology have important roles in determining our material well-being and in shaping society
- prioritisation of research efforts is influenced by economic, social and ethical considerations, as well as by scientific ones
- the earth has finite reserves of fossil fuels and other resources
- economic considerations are not always paramount in determining the location and the mode of operation of industry
- pollution is a major problem in industrialised societies.

The second stage in the assessment scheme design process is to consider the kinds of intellectual skills that the curriculum has been designed to develop. Bloom and his co-workers (1956) grouped these intellectual skills into a hierarchy of six categories: knowledge (or recall), comprehension, application, analysis, synthesis, and evaluation, as discussed in chapter 2. Of course, Bloom's is not the only taxonomy of cognitive activities. For example, Ebel (1972) has produced a seven-category hierarchy, which is somewhat similar to, but rather simpler than, Bloom's, and a number of Examination Boards have adapted the Bloomian scheme to suit their particular purposes. We have no reasons for promoting Bloom's categories rather than any others, but we do urge the adoption of some category system, however simple. A category system acts as an *aide-mémoire* in constructing tests, so that we avoid the trap into which so many teachers fall of basing all test items on the ability to recall, thereby trivialising the test. The Bloom taxonomy has too many inconsistencies and incoherences to be the answer to all our assessment problems, but it does serve to remind us that we should be teaching, and assessing, higher-order intellectual skills.

As far as less able children are concerned, we believe that it is realistic to engage in application – that is, the ability to use learned material (concepts, methods, principles, laws, and theories) in new and concrete situations and to use general principles to solve specific problems. Provided that the context is appropriate and well chosen, we believe that such children can engage in analysis – that is, break down material into its component parts so that its organisational structure may be identified and understood, and in synthesis – the ability to put parts together to form a new whole. In this latter category we would place activities such as experimental design, the writing of an essay on a topic, the development of a classification system, etc. There is no reason why such children should not engage in evaluation, the highest level skill of all – that is, the ability to judge the value of material for a given purpose. In the context of science education this could be an activity such as the criticism of an experimental design or of the conclusions drawn from experimental data, or the evaluation of a theory in terms of its observational support. All of these activities are within the capabilities of less able children, provided proper consideration is given to the context, the conceptual demand, the linguistic complexity, and the relevance to the lives of the children. What is most certainly true is that, if such activities are part of our curriculum targets and part of the curriculum experiences of children, they should be part of the assessment scheme.

We recommend the adoption of a test specification that identifies, in advance, the distribution of test items between the various intellectual skills. Using a planning matrix, such as that in figure 7.7, it should be possible to ensure that a test adequately measures the range of intellectual skills we seek to develop in less able children.

At the very top of any hierarchy of intellectual activities we would place creativity. Scientific creativity is a goal we seek to attain in our curriculum – not least because of its beneficial effect on interest and motivation – and it should, therefore, be represented in our assessment scheme. However, it is an elusive concept. It is easier to

Content items (from figure 7.6)	Recall	Compre-hension	Applic-ation	Analysis	Synthesis	Evalu-ation
Concepts and terminology						
Facts						
Trends and sequences						
Classifications						
Principles and generalisations						
Theories						
Experimental techniques						

Figure 7.7 Test planning matrix (second step)

recognise it than it is to describe, promote and measure! It is an unfortunate consequence of many teachers' concern to teach correct knowledge (in Kuhn's terms, the accepted paradigm) and to eliminate misconceptions in children's understanding of science that they may, unconsciously, tend to discourage speculative thought in children and thereby foster the mistaken view that science is intolerant of individual opinion. If children are to understand the creative aspects of scientific practice they must be provided with opportunities to think creatively.

Medawar (1969) identifies four kinds of scientific creativity: deductive intuition, inductive intuition, experimental design, and instant apprehension of analogy. Children should be provided with opportunities to develop and practise each of these skills, and we must make an effort to assess their progress in them. Technology may well be a better vehicle than science for fostering creativity, because technological problems never have a single, correct solution. Project work also provides the opportunity for children to use their knowledge and process skills in idiosyncratic, personal, and creative ways. Assessment of project work and problem-solving activities in technology, together with test items designed to assess Medawar's four categories of creativity, should provide adequate test data in this area. Use may also be made of the kinds of test items produced by Torrance (1974) and by the Assessment of Performance Unit (APU, 1985).

The next stage of the assessment procedure is the selection of assessment technique, writing or assembling the items and constructing the tests. The authors of Examination Bulletin 27 remind us just how important it is to use well-designed assessment instruments:

> No assessment can be better than the examining instruments allow. If the instruments are faulty, the assessment will be faulty. This may appear to be a truism, but examination results are so often taken at their face value that it needs to be stated. It is proper, therefore, that the instruments of assessment must themselves be subject to evaluation. (Schools Council, 1973a)

The following general criteria of evaluation are suggested:

- the instruments of assessment should be valid
- they should be reliable
- they should not significantly distort acceptable educational practice
- they should not make unreasonable demands on manpower, finance, time, or facilities.

There are, of course, many ways of classifying assessment items. Mathews (1972) suggests that the main criterion should be 'the degree to which the pupil can respond as he pleases and still gain credit for doing so'. Thus, there are two extreme types:

1. The objective or fixed response type, to which candidates must respond exactly as the item writer predetermines or achieve no mark (they may even incur a penalty). The score of a particular candidate on a test of objective items is wholly independent of the marker. Examples include true/false, multiple-choice, multiple-completion, assertion/reason and related types. Apart from the first two styles, these often present linguistic and procedural complexities that render them unsuitable for underachievers.
2. The free response type, to which there is no predetermined answer, and in response to which candidates select the material they deem most appropriate.

Some of the advantages and disadvantages of these extreme types are listed in figure 7.8. Between the two extremes lie other types, such as the structured questions now much used in GCE and CSE science examinations. Additional types that are particularly appropriate to slow learners include completion items, in which the candidate supplies an answer that is only a few words or sentences long, and 'restricted response essays', in which the pupil lists, for example, the observations or stages in a particular experiment, without having to construct lengthy paragraphs. We also advocate the use of diagram completion items. Rather than requiring children to select appropriate apparatus for an experiment and then to draw and label a suitable diagram, an activity that many less able children would find extremely difficult, we can present an unlabelled diagram for completion or a fully labelled diagram for which they ascribe a use. In other words, we are advocating the kinds of DARTs activities identified in chapter 5 (pp. 88–92) as central to the learning experiences of less able children. In making this suggestion, we are reinforcing our view that, whenever possible, the gathering of assessment data is based on usual classroom activities. We wish to avoid the stress that often accompanies an 'official' test or examination.

What is abundantly clear is that assessing a range of intellectual skills, practised on a range of content, requires a range of assessment styles and methods. Detailed advice on the choice of appropriate assessment techniques in science can be found in recent works by Harlen (1983) and Kempa (1986). It is also clear that a curriculum that provides for children to follow a variety of learning routes, and to engage in a degree of negotiated content, will need to provide an assessment scheme suited to individual needs. This

	Free Response	Objective
Abilities measured	Requires the candidate to express himself in his own words, using information from his own background and knowledge.	Requires the candidate to select correct answers from given options.
	Can test high levels of reasoning – inference, organisation of ideas, comparisons, etc.	Can test high levels of reasoning – inference, organisation of ideas, comparisons, etc.
	Does *not* measure purely factual information efficiently.	Measures knowledge of facts very efficiently.
Coverage of syllabus	Covers only a limited field of knowledge in any one test. Essay questions take so long to answer that relatively few can be answered in a given period of time. The candidate who is especially fluent can often avoid discussing points of which he is unsure.	Covers a broad field of knowledge in a single test. Since objective questions can be answered quickly, one test may contain many questions. A broad coverage provides reliable measurement.
Motivation	Encourages pupils to learn how to organise their own ideas and express them effectively.	Encourages pupils to build up a broad background of knowledge and abilities.
Ease of preparation	Requires only a few questions per test. Tasks must be clearly defined – general enough to offer sufficient scope, specific enough to set limits.	Requires many questions for a test. Wording must avoid ambiguities and 'give-aways'. Distractors should embody the most common misconceptions.
Marking	Usually very time consuming to mark.	Can be marked quickly.
	Permits teachers to comment directly on the reasoning processes of individuals. An answer may be marked differently by different markers or by the same marker at different times.	Answer generally scored only right or wrong, but scoring is very accurate and consistent.
Choice	Usually candidates are allowed a choice of questions.	Choice usually not allowed.

Figure 7.8 Comparison of free response and objective questions

seems to lead directly to the notion of profiling, an issue we shall discuss in chapter 8 (pp. 187–91).

Processes and attitudes

The principles of test construction outlined above are just as applicable to the assessment of processes and attitudes as they are to the assessment of knowledge and its manipulation, and so will not be reiterated.

The processes to be assessed are those listed P1–P21 in table 3.1. At first glance, it would seem that cognitive processes are best tested by means of written tests and that psychomotor processes are best tested by means of practical tests. Closer examination reveals that, as far as the processes of science are concerned, it is not always possible to distinguish the cognitive from the psychomotor. Consequently, the assessment of a pupil's level of competence in the processes of science requires a combination of observation, scrutiny of laboratory notebooks, interviews, and tests. If observation of laboratory behaviour is to be systematic and informative, then it may be necessary to employ checklists (e.g. 'notices important features of objects', 'identifies similarities and differ- ences', 'rechecks observations', etc.). Processes such as P6 (selec- tion of appropriate measuring instruments) can be assessed by means of short practical tests, or by pen and paper tests (Bryce *et al.*, 1983). Processes requiring the following of instructions, the use of manipulative skills, or the application of procedural skills are best assessed by teacher observation during the practical work, or by examination of the product at the end of the activity.

Because each skill has to be assessed several times to ensure reliability, the testing and marking load is likely to be extremely heavy. Therefore, it is sensible, whenever possible, to build the assessment activity into the normal classroom learning activity. This is exactly the situation that will arise with the introduction of the extensive practical testing associated with GCSE. Such an approach, of course, requires extensive forward planning and very careful record keeping. Nothwithstanding the increased availability of microcomputing facilities in schools, which offer some prospect of more efficient and economical data processing and record keeping, the implications for teacher support and teaching loads are considerable.

Mayer and Richmond (1982) provide a comprehensive overview of assessment instruments in science education, including some dozen concerned with understanding the nature of science, and 32 for assessing science process skills. This latter area is the subject of much recent research activity in the USA, and the attention of readers is drawn in particular to the Test of Integrated Process Skills (TIPS), designed by Okey and his co-workers (Burns, Okey and Wise, 1985; Dillashaw and Okey, 1980). Nor should we neglect the valuable work carried out by the Assessment of Performance Unit and published in its reports *Science in Schools at Ages 11, 13 and 15*, from 1981 onwards. Lists of manipulative/laboratory skills appropriate to school science can be found in Guthrie (1980), and valuable advice on the testing of such skills is contained in the Techniques for the Assessment of Practical Skills (TAPS) scheme (Bryce *et al.*, 1983) and in the work of Sands and Bishop (1984). The TAPS project, in particular, has produced well over 300 items of assessment that, taken together, form a comprehensive pupil profile.

By way of illustration, consider a sequence of items relating to objective P15: 'The pupil should be able to use a compound microscope' (McCall, Bryce and Robertson, 1983). Note how the test items reflect a progressive growth in competence by the pupil, and how the pacing of the work can be adjusted to suit the differing rates of progress of children. Any one test item can be repeated with different materials, should a child fail the test on any occasion. With the advent of GCSE this will become an important facility. Incidentally, the TAPS package provides all the materials necessary for completing the tests. Teachers are not expected to photo-reduce or make acetate sheet slides.

Test item P15.1   Microscope labelling.
Small lettered cards are attached with Blu-Tack to the type of microscope normally used in class. The pupil is required to match parts as indicated by the letter with functions by writing in the appropriate letter e.g.:

This is where you put your eye
This is where the slide is put

Test item P15.2   Secret signs.
Simple geometric symbols are photo-reduced on to acetate sheet and mounted singly on microscope slides. Each symbol is circled with marker pen so that under low power magnifications a coloured circle is visible, even though the symbol is out of focus. The pupil is required to find the middle of the circle, then focus the microscope to find the 'secret sign'. It ought to be possible for class sets of slides to be available. By varying the colours of the circles and symbols, the teacher has an easy method for scoring responses.

Test item P15.3   Slide search.
Five symbols are mounted inside a large circle. One of the symbols is circled in colour as in P15.2 above. Having focussed on that, the pupil is required to move the slide without altering the focus to find the other signs within the large circle.

Test item P15.4   Focussing sandwich.
Sets of microscope slides have been prepared by mounting two
photo-reduced letters on acetate sheet, one above the other in 'the
middle' of the slides. The pupil has to find both letters on each slide.
As the coloured circle aid is missing on this item, the pupil must
display competence in both focussing and slide searching.

In chapter 2 we discussed the priority afforded to motivation in
our curriculum proposals. We argued that good motivation is a
prerequisite of academic achievement. However, we believe that as
far as assessment and evaluation are concerned it can also be
regarded as an achievement in its own right. Many of the pupils
who are our immediate concern in this book come to school, or
develop in school, very low motivational levels, largely because the
curriculum and the 'climate' (namely, too little individualisation,
too little experience of success, too little experience of 'being in
control') are inappropriate to their needs and aspirations. There is
no doubt that a reduction in motivational levels as a consequence of
curriculum experiences – and its associated adverse effect on
attainment – would be regarded as a legitimate criticism of a school
science department. If the school succeeds in generating, or
regenerating, good motivation, such that it leads to improved levels
of attainment (as we have argued it would), then it can be claimed as
a major achievement on the part of the school and the individual.
Consequently, the assessment of motivational levels should count
as part of the curriculum evaluation procedure. However, we are
not proposing that they be formally assessed as part of the
assessment strategy. But we are proposing that other affective
characteristics (coded A1–A17 in table 3.2) are made part of this
procedure.

Assessment of attitudes is more difficult and less reliable than the
assessment of cognitive and psychomotor skills. It is an area that is
largely unfamiliar to practising science teachers, using, as it does,
techniques developed in psychology and sociology. These techniques include observational studies of children, questionnaires,
interviews, and project work. It is not possible to discuss these
techniques within the constraints of this chapter, save to note that it
is clear that, for the effective assessment of a course in which the
development of attitudes is a major goal, we need to employ a
variety of methods, and that self-reporting should comprise a
significant part of the procedure. Henerson, Morris and Fitzgibbon
(1978) provide detailed guidance on the design of questionnaires
and the development of interview and observation procedures;
Sutton (1986) gives much helpful advice on the construction of pupil
self-assessment methods; Noll, Scannell and Craig (1979) provide a

rigorous treatment of the main principles underpinning attitude assessment.

Interpretation and reporting

Once test data have been obtained they have to be interpreted. A score of 60 per cent on a physics test is meaningless without some reference point from which to judge whether it is a cause for rejoicing or despair! In principle, there are three bases of interpretation.

- comparison with some predetermined standard of performance (criterion referenced)
- comparison with the performance of others (norm referenced)
- comparison with the individual's previous performance (self referenced).

Self-referenced assessment is central to the proposal, in chapter 2, that the priority in curriculum design should be afforded to motivation. Successful motivation depends, crucially, on individuals recognising that they are making progress. To a large extent it does not matter how well or how badly other children are progressing. In a *norm-referenced* scheme, however, the performance of other learners does matter. Indeed, the score of each individual is related to the scores of others – ideally to those individuals matched in terms of age and background. The test designer starts from the assumption that individuals vary in their levels of knowledge and skills, in their proficiency in various tasks and in their possession of certain characteristics, and attempts to maximise these differences in order to differentiate individuals as sharply as possible. The aim is to identify the norm, or average performance, and to describe each individual's performance in terms of differences (upwards or downwards) from the norm. In our view, such assessment is primarily the responsibility of the examination boards (though GCSE proposals seek to change the basis from norm referencing to criterion referencing) and has no place in the curriculum for less able children until such time as preparation for external examinations becomes a dominant consideration – and, clearly, such is not the case in a curriculum designed to achieve scientific literacy for all. For such a curriculum, it is more appropriate to employ *criterion-referenced* assessment. In this case an individual's score has meaning without comparison with the performance of others. This time we are interested in finding out whether a learner has reached the particular goal we had in mind (the criterion) or not. Questions in the test are designed to sample various aspects of the criterion with the purpose of discovering the level of mastery of each individual and, as a consequence, of the whole learning group.

A typical criterion-referenced test is the MOT driving test. Each candidate must pass on each performance objective (vehicle control, reversing, parking, signals, highway code, etc.). The appropriate standard must be attained by each candidate if they are to pass. What other candidates score has no bearing on whether a particular person passes or fails. Criterion-referenced assessment is appropriate to the measurement of levels of attainment, to the identification of learners experiencing difficulties, and to the evaluation of learning experiences. Detailed technical advice on the construction of criterion-referenced tests can be found in recent works by Brown (1981) and Black and Dockrell (1984). Some of the problems peculiar to criterion-referenced testing in science are discussed by Kellington and Mitchell (1980), Stillman (1982) and Johnstone *et al*. (1983).

Several points need to be made at this juncture:

1. Analysis of test results is an essential part of the assessment and evaluation process. Unfortunately, many teachers set elaborate tests and then fail to analyse the results, thus devaluing their initial efforts (Hodson, 1986b).
2. We need to set, in advance, the target performance that we will accept as evidence of adequate mastery. In the next chapter we shall argue that the ideal organisation for the operation of a curriculum for universal scientific literacy is individualised, resource-based learning. Such a system requires teachers to adopt an extensive detailed recording system, capable of giving instant access to details of pupil experiences and attainments. As far as criterion-referenced tests are concerned, it is simply a matter of recording a tick in the appropriate box of the record card when an acceptable level of attainment has been obtained (figure 7.9). Scrutiny of the record cards would quickly reveal that certain children (numbers 6 and 10, for example) are experiencing learning difficulties and require help of some kind. It would also be apparent that the curriculum is functioning adequately, for most children, on the topics covered by tests A and D, whilst there is some cause for concern regarding the curriculum units covered by tests B and C.
3. We need to have confidence in our test items. Often, teachers fail to make a decision about 'what is on trial' when they set a test. They are unsure whether to assume that the test items are 'good', so that a low score indicates inadequate teaching materials, whether to assume that the curriculum is 'good', so that a low score indicates poor assessment items, or whether to

| Pupils | Tests | | | | | % Pass |
	A	B	C	D	etc.	
1	✓	✓	✓	✓		100
2	✓	✓		✓		75
3	✓	✓		✓		75
4	✓	✓		✓		75
5	✓	✓		✓		75
6						0
7	✓	✓		✓		75
8	✓			✓		50
9	✓	✓		✓		75
10						0
11				✓		25
12	✓	✓		✓		75
13				✓		25
14	✓			✓		50
15	✓			✓		50
16		✓	✓	✓		75
17	✓			✓		50
18	✓			✓		50
19	✓			✓		50
20	✓	✓		✓		75
% Pass	75	50	10	90		

Figure 7.9 Test data and their interpretation

assume that both are 'good' and the children are on trial. Unless we know where we can place our confidence, we cannot know how to interpret the test data.

There would seem to be a tendency on the part of many teachers to assume that a curriculum package produced by teachers, for teachers, must be sound. The teachers observed during the Manchester Assessment Project (Hodson, 1986e) tended to assume that low scores indicated inadequate assessment instruments, rather than inadequate learning experiences. The development of valid and reliable test items, in which teachers have confidence, is an urgent priority if teachers are ever to be convinced to turn their attention to critical and informed evaluation of curriculum materials and teaching methods.

Writing good, reliable test items is a skilled occupation, requiring extensive practice. It is more economical in the long run to compile a bank of well-tested items, with clear specification of intellectual skill and subject matter tested, than to train all teachers in the skills of item writing. In this respect the situation is similar to that attending computer programming. It is an important current task to acquaint all teachers with the capabilities and potential of computer-assisted learning and computer-managed learning. However, it would seem to be more economical if time, energy and resources are concentrated in one or two individuals, and their expertise made available to others on a consultancy basis, than attempts are made to furnish all teachers with such expertise. In the same way, all teachers should be made aware of the scope and possibilities of systematic assessment and evaluation, but test design skills need not be given to everyone, provided that such skills and the products of those skills be made available to all teachers via a centralised LEA consultancy service and an item bank.

How should we respond?

Basically, there are two responses: the reporting response (the supply of information to whomsoever needs it) and the teaching response (the implementation of courses of action appropriate to the strengths and weaknesses diagnosed in individual learners). As far as reporting is concerned, there are at least four targets: the learners themselves, parents, other teachers, and employers (which, in this context, includes institutions of further education). We do not have the opportunity here to enter into a detailed discussion of the relative merits of percentage scores, grades, class positions, and so on. Instead, we confine ourselves to the observation that many recording and reporting systems currently employed in schools conflate detailed information into almost meaningless grades and scores, and thereby lose whatever useful assessment data has been collected. Clearly, we need a recording system capable of storing all the information we have collected and a reporting system that provides the recipient with sufficient information, in an appropriate form. Clearly, too, that information and that style of reporting depend on the reporting target. It was stated earlier that higher test scores are obtained when courses employ frequent testing. This enhanced performance is even more marked when feedback to the learners is rapid and informative. Schools should give more attention to the way they respond to assessment data, making feedback to the pupils a major priority. There is no doubt that the contemporary trend is towards more extensive and elaborate reporting to parents and employers, in the form of profiling. We wholeheartedly support this trend, but would urge teachers to proceed cautiously. It is a cardinal error to seek to report information about pupils that we have not got and do not know how to obtain. Some of the more avoidable pitfalls are described by Law (1984), and some teacher anxieties regarding profiling are discussed by Hodson (1985b).

In summary, we are arguing that an effective assessment system should have a number of characteristics:

- it must provide a comprehensive description of each child's attainment in several areas: knowledge, processes and attitudes
- it should relate those attainments to the past performance of the child, to the course objectives, and, possibly later in the course, to the performance of other children
- it should clearly identify strengths, weaknesses and problems in all these areas, and attempt to identify possible sources of low attainment
- it should provide a continuing record of every child throughout the period of secondary schooling and provide regular, detailed feedback to the child, the parent and other teachers.

Viewed in this way, assessment becomes an integral part of teaching, guiding the pupil through the curriculum, assisting the teacher in making informed judgements about the most appropriate curricular action and provision for each child, and providing detailed information about the child's strengths and weaknesses, in the form of a profile of experiences and attainments. The following chapter includes a discussion of how such a system might be organised.

8

Content and organisation

INTRODUCTION

There is today still a degree of controversy regarding the principles that should inform and guide the selection of content of a science curriculum. Black and Ogborn (1981), whilst not addressing themselves directly to the problem of science for underachievers, believe that special problems deserve special solutions.

> The – in our view – deeply serious efforts made within schools to value all pupils equally and to respect their qualities in ways not restricted to academic ability, have badly frightened or upset those in our society who – quite correctly – see that valuing all does in the end mean not valuing some specifically, and who do not want to give up that special valuation.

The implication of this statement is that special children demand, and are entitled to receive, special consideration. This is particularly true of underachievers who, usually through no fault of their own, are failing to capitalise on an education that does not value them sufficiently.

There is no nationally prescribed content for a science course for low achievers in England and Wales, and it is certainly not the aim of this chapter to provide a specific proposal. Indeed, to do so would serve to invalidate much of what is written elsewhere in this book, not least about the responsibility that individual science teachers must take upon themselves in the selection of content appropriate to the perceived needs of individual children in particular schools at a particular time. Rather, what we shall do is to rehearse the principles that should guide the selection of content. Some of these principles are immutable, such as the need to select content upon the aims and objectives produced for the course and the need for individualisation. Others may vary over time, for example, as societal needs are perceived to change, or new educational perspectives take priority. In the latter case, selection of content will be guided by the gurus of the time, be they authoritative practitioners, administrators, academics, or professional bodies such as Her Majesty's Inspectorate (HMI), the Assessment of

Performance Unit (APU), the Association for Science Education (ASE), and the Secondary Science Curriculum Review (SSCR), for all have made important contributions to recent debate on appropriate science content. Much of the work of the professional examining bodies, such as the Secondary Examinations Council (SEC) and the Northern Examining Association (NEA), has recently concentrated on course content for the new General Certificate of Secondary Education (GCSE) examinations. Their experiences need to be tapped, particularly since the GCSE is conceived as an examination for all children and, for the first time, brings underachieving children under the 16+ examination umbrella – a move, incidentally, that runs counter to practice in almost every major advanced country in the western hemisphere, including the United States of America, Scandinavia and most other European countries, Australia and, most recently, New Zealand.

The problem of appropriate science content for underachievers may also be seen in multicultural terms. Such an approach to content will need to attach importance to the political, ethical, and cultural dimensions of a pluralistic society. Traditionally, politics has not imposed itself upon the science curriculum. However, an increasing awareness of the importance of the science and society interface has come about as a result of such tragedies as the thalidomide scare, the Chernobyl nuclear power station disaster and the influence of new technologies on employment and lifestyles generally. A disproportionately large number of underachievers have non-white ethnic origins, and their awareness of science may be dulled by the use of inappropriate exemplars in school or, as is argued later, a philosophical assumption about the nature of western science.

Of course, the most fundamental sub-cultural division is by gender. Much recent research has identified the general underachievement of girls in the physical sciences and, to a lesser extent, the underachievement of boys in the biological and human sciences (HMI, 1980; Kelly, 1981a, 1985; Harding, 1983; Walford, 1983; Smail and Kelly, 1984a,b). Whilst there have been a number of intervention strategies for increasing the representation and raising the attainment levels of girls (Kelly, Whyte and Smail, 1984; Ferry, 1985; Smail, 1985), there seems to have been little attempt to institute similar strategies on behalf of boys. We would urge that attention is directed to combating underachievement by all sub-groups.

PRINCIPLES UNDERLYING CONTENT SELECTION

In discussing the nature of the proposed content, it is important to be alerted to the disease of 'curriculum conditioning', whereby content is justified by the retrospective articulation of aims and objectives. It is easy for those of us who have benefited so greatly from the content of our own courses to assume that the same kind of science will be beneficial to others. It is worth repeating, therefore, that we envisage three major long-term aims of the science curriculum for low achievers, presented here in descending order of importance.

1. *Learner-centred aims*
   These aims concern motivation, the development of attitudes and feelings, the enhancement of self-image, and an improvement in overall lifestyle, in which science can play a part. The content most apposite in this context will be that which encourages a questioning curiosity about natural phenomena. It is here that method, organisation, and content become inextricably fused. Project work for example, so successful a component of the science curriculum for the low achievers at Castle Hill school (see chapter 9, pp. 196–9), is almost independent of content.
2. *Society-centred aims*
   These have been described as aims concerned with the interface between science and society, and will involve moral and aesthetic considerations. Technology-oriented science is of particular importance under this heading, and content such as nuclear energy provision, population control, and pollution is particularly relevant.
3. *Science-centred aims*
   These aims are related specifically to the attainment of scientific knowledge, concepts, and theories. They are concerned with scientific 'method', and they are the essential prerequisite to scientific literacy.

The fundamental principle to be followed in the selection of content is that it should be able to motivate the child. We are certainly not alone in stressing this (Peck & Williams, 1978; Clegg & Morley, 1980; Dark *et al.*, 1985). Enthusiasm for science and for learning is likely to be stimulated by the following:

- relating content to real life situations
- using children's own experiences and knowledge as starting points
- using children's own interests as starting points
- involving children directly in experimentation and other active learning methods
- choosing topics that can focus on societal conerns, such as energy resources and world food production

- giving a sense of personal satisfaction and self-esteem through successful learning
- affording children a measure of choice and self-determination through the practice of negotiated content
- providing a wide variety of learning experiences, stimuli and tasks
- developing skills that are seen to have direct relevance to future work and leisure activities.

This is not to say that content *per se* is to be regarded as unimportant. We recognise that the concept of scientific literacy implies a certain knowledge content, and we regard certain facts, concepts, and theories as worthwhile knowledge. However, provided the teacher covers a certain minimum content, the major criterion for the selection of knowledge content is that it should relate positively to the curriculum objectives concerning processes (table 3.1, P1–P21) and attitudes (table 3.2, A1–A16). Of course, processes cannot be learned independently of content – one cannot make accurate observations (P5) unless one has decided what to observe and how to observe it. This may also raise questions about the choice of instrumentation (P6), measurement (P7), and style of communication (P8). It quickly becomes apparent that the phenomena or objects we choose to observe (the observation content) determine the precise nature of the process skills acquired. The content should, in addition, have social meaning and usefulness, enhance and improve each pupil's life as a citizen and family member, provide a functional understanding of the natural world, and ensure an awareness of contemporary science and technology and its interaction with society. Thus it is important that teachers monitor the content negotiated by the children in order to ensure adequate coverage in both breadth and depth. What we seek is a judicious blend of free pupil choice, negotiated selection, and teacher direction. This approach has profound implications for organisation at the departmental level, which in turn accounts for the dual nature of this particular chapter heading.

Ensuring success is not easy with underachievers, particularly when at the same time the success must be seen as worthwhile. What we can do is to set short-term, achievable objectives, and provide regular and detailed feedback on the extent of learning achieved by each child. Thus, we envisage a curriculum organisation based on successive levels of attainment, through which children acquire an escalating series of school awards (certificates of attainment) on a test-when-ready principle. This notion was explored in chapter 7 (p. 146), and examples of it in action are to be found in figure 9.4 (p. 204).

We also envisage a modular, or topic, approach, largely because it would offer the flexibility we seek – to allow for several different learning routes, to accommodate individual differences in aptitude, linguistic skills and preferred learning style, and to permit a measure of negotiation of content.

EXEMPLARS OF CONTENT

We have emphasised several times that the final selection of content should be the responsibility of the individual teachers, or groups of teachers, who are best placed to take into account local conditions, including their own interests. However, if the global aim of our course is to be universal scientific literacy, there are some topics and issues that we feel ought to be included. In order to highlight the need to relate content to the real world of the child, an indication is given in parentheses of what content is more applicable. The following topics might form the basis of *core modules*:

1. Understanding ourselves: body and mind (health, personal relationships).
2. The nature and structure of the universe (Halley's comet, Strategic Defence Initiative, astrology).
3. Life on earth, and the interaction of life forms with their environment (human evolution and the development of race, parasites).
4. Natural resources (diamonds, coal, oil, water).
5. Energy and energy resources (including nuclear energy, energy in industry, home and commerce).
6. Population and pollution (birth control, Chernobyl).
7. Food production ('fast' foods versus 'real' foods).
8. Synthetic materials (clothes, plastics).

The basic course will be reinforced with 'special interest' modules, to be negoiated between teacher and pupil. The extent to which free choice is allowed will depend, of course, on the particular situation in the school (staffing, resources, cost, etc.). At a later stage, it is to be hoped that some of the negotiable modules would include a number specifically designed as GCSE preparation modules. The following list gives examples of the sorts of things we have in mind:

Photography	Electronics
Meteorology	Colour
Brewing and wine making	Building science
Cosmetics	Flight, natural and powered

Horticulture	Radioactivity
Forensic Science	Science and cooking
Metallurgy	Great scientists
Science and sport	Explosives
Biotechnology	Kinematics and dynamics
Animal behaviour	Radio, cassettes and TV
Communication	Transport
Hospitals	Age

A list is required that is sufficiently lengthy and versatile to cater for all ranges of ability and interest. The Northern Examining Association (NEA, 1986) has comprehensive proposals for a modular science GCSE syllabus, to be examined for the first time in 1988. In this it suggests four core modules, followed by a series of 'option module' titles. This reflects the model proposed here, although the individual modules in our proposal are shorter and of wider variety, in order to cope with the shorter concentration span of underachieving children. Readers are recommended to refer to this document for details of content within the modules.

In order to illustrate how the principles we have been advocating throughout this book can be implemented in the design of curriculum modules, we have sketched an outline of Module 1 (Understanding ourselves; body and mind) in table 8.1.

The motivational aspects of the course are apparent in the choice of starting and finishing points. The programme starts by inviting children to learn more about themselves and to compile personal profiles. There is ample scope here for measurements and for comparisons, leading to the construction of data tables and bar charts, as children come to recognise the superficial differences between sex and race. The module finishes with a sub-unit on elementary first aid. There is a possibility of developing these skills to the point at which children could be certificated by the British Red Cross Society or the St John's Ambulance Service.

It is also important to establish clear links with other modules, lest the course degenerates into a ragbag of assorted oddments. Obvious points of contact are between unit 1.7 (sex education) and Module 6 (population and pollution), and between unit 1.3 (food for energy and growth) and Module 7 (food production). However, the criterion of scientific literacy itself acts as a unifier. In all these modules, the skills and processes the children will have to demonstrate in order to obtain their certificates of achievement will not be dissimilar (see, for example, figure 9.4). The NEA has proposed that, in addition to core and negotiated (which it refers to as 'optional') modules, a child might also undertake a 'unifying' module. Such a module is 'designed to provide candidates with the

Table 8.1 *Units of Module 1, 'Understanding ourselves: body and mind'*

Unit	Focus of attention
1.1 Me and my body	Personal facts (photograph; height, weight, sex). Skin, hair; colour and texture. Comparison with others; with and between races.
1.2 Keeping healthy	Need for exercise. Body tone v. fitness. Dental care. Effects of alcohol, drugs, tobacco. Liver, heart and lung problems.
1.3 Food for energy and growth	Need for energy; well-balanced diet, including roughage. Food additives. How to read food labels. Effects on body of starvation v. obesity.
1.4 Respiration	Oxygen and carbon dioxide; heart and lungs. The chest.
1.5 Skeleton and muscles	Size, support and strength. Efficiency in terms of speed and weight. Forces, levers; tendons and ligaments.
1.6 Our senses	The five senses; sensitivity, interactions, efficiency. The sixth sense; ESP, sleep, hypnosis.
1.7 Sex education and parenthood	Growth and development in and out of the womb. Contraception. Child care; the family as a nucleus; alternatives – adoption, surrogacy.
1.8 Medical care and disease	Body defences; germs; immunisation. Keeping warm, dry, clean. Infectious diseases. Analgesics and patent medicines. Hospitals; X-rays; stethoscopes. Elementary first aid.

opportunity to display the range of skills and abilities they have developed during the course'. The idea of a unifying module is an interesting one, although it is clearly perceived by the NEA as an aid to assessment, and as such might be more accurately described as a 'revision' module. Our unifying module would be specifically designed to select areas of practical and theoretical interfacing.

As far as specific content is concerned, we are anxious to ensure that those who will cover only the basic minimum on any particular module will not be overloaded with conceptual complexity and abstraction. Thus, for example, unit 1.3 concerning nutrition might classify foods into four groups – meat and fish; cereals and bread; fruit and vegetables; milk and dairy produce – and suggest that a balanced diet requires items from each group, and that food production techniques are concerned to maximise output in these four areas. Extension sub-units, for higher-ability children, would deal with the same issues in terms of protein, carbohydrate, vitamins, minerals and fats. The principle we are trying to establish is an operational/functional understanding in the first instance, with abstract/theoretical knowledge coming afterwards, where possible. For each unit in the module it is necessary to identify the required learning content, both for the basic minimum provision and for any extension activities. Each of these 'knowledge items' could be coded (K1 to K*n*, see table 8.2).

Even quite straightforward relationships between science content (for instance, K5 and K6 in table 8.2), which are so obvious to teachers that they never think of drawing special attention to them, may go unnoticed by less able and underachieving children. Such children become confused by overstimulation, and have to be taught how to extract the relevant from the irrelevant, and how to put the relevant back together in some sort of pattern that makes sense to them. This is why it is essential to give them a head start by using patterns with which they are already familiar in their everyday lives, whilst at the same time taking care to stress the differences between everyday and scientific knowledge. It is not surprising that this should be so, since Bloom (1956) places both analysis and synthesis high up in his hierarchical taxonomy. A unifying module would set out specifically to identify some of these implicit relationships, for example between communication techniques and biotechnology, and between overpopulation and pollution.

POLITICAL, ETHNIC AND GENDER INFLUENCES ON CURRICULAR CONTENT

The production of national criteria governing all GCSE syllabuses is a major educational innovation. These criteria 'represent a consensus among the teaching profession, the examining boards, the users of examination results, the secondary examinations council [SEC] and the Government' (SEC, 1986). In spite of the dominance of the GCSE as an examination procedure, they nevertheless permit the development of different syllabuses within agreed guidelines. In particular, the general criterion that gives guidance on bias is of significance when dealing with underachieving children.

Recognition of cultural diversity
In devising syllabuses and setting question papers, Examining

Table 8.2 *Module 1: 'Understanding ourselves: body and mind'. Focus on 'K', 'P' and 'A' components and associated activities*

Unit	Knowledge 'K'	Processes 'P'	Attitudes 'A'	Some possible activities
1.1 Me and my body	K1 Height and weight are age and sex related K2 Variation within and between individuals and races K3 Continuous variation (skin hue) K4 Eye colour and genetic inheritance also K5 Hair colour and texture K6 Weight is culturally influenced	P5 Accurate observation P6 Selection of appropriate measuring instruments P7 Accurate measurement P8 Describing and reporting P14 Processing data P15 Presenting data P21 Formulating hypotheses	A4 Intellectual curiosity A5 Tolerance of others A7 Co-operation with others A8 Open-mindedness	Developing own instruments, e.g. colour cards for comparing eye and skin colours reliably Photographing against a scale Making a family tree (of eye/hair colour) Relationships between sets of data on 1:1 basis, e.g. bar charts, pie charts of sex and height and weight Also race and skin hue
1.2 Keeping healthy	K1 Teeth numbers K2 Type and function K3 Caries K*n* etc.			

———————— and so on ————————

Groups should bear in mind the linguistic and cultural diversity of society. The value to all candidates of incorporating material which reflects this diversity should be recognised.

Avoidance of bias
Every possible effort must be made to ensure that syllabuses and examinations are free of political, ethnic, gender and other forms of bias. (Quoted in SEC, 1986)

Selection of content appropriate to underachievers should take into account the three caveats of politics, ethnic origin, and gender. We shall pay little attention to the first of these, since current concerns about political bias in the school curriculum (e.g. peace studies) do not appear to impinge too heavily upon the teaching of science to underachievers. The teaching of scientific fact, which may or may not appear unpalatable to politicians, is not our first concern in this book in any case. Ethnic origin and gender are, however, very relevant.

Ethnic bias and the content of the science curriculum

HMI's survey of underachievers in secondary schools (HMI, 1984) indicates that the proportion of children from ethnic minority communities varies from none at all to over 50 per cent in some schools. Science educators are only just beginning to think seriously about the contribution that science teaching makes to the sustaining of racist stereotyping and racism, and they are not alone in this respect. *The School Curriculum*, (DES, 1981a) lists as one of its overall aims for schools 'to instil ... tolerance of other races, religions and ways of life', and this is stressed again in the 1985 document, *The Curriculum from 5 to 16* (DES, 1985c).

Little positive advice is given as to how this might be accomplished through science education. The question is whether tolerance of ethnic minorities is, of itself, sufficient, or whether curricular materials need to be devised that reflect positive anti-racist ideology. Pervading much of the science taught in British schools is the philosophical assumption of the superiority of western science and western scientists, which in turn implies a devaluation of non-western cultures. It is evident from our discussion about the personal nature of scientific understanding (chapter 2, pp. 36–8) that there are different legitimate stances from which scientific 'fact' may be interpreted, and that for science education this means attempting to place the learning of science in a more personal context. Overpopulation may appear to an indigenous white person living in England to be a major cause of the problems facing certain ethnic groups, in terms of poverty, disease, and starvation. Yet the cultural traditions and religious beliefs of societies that depend upon large families to supply the needs of the elderly permit a quite different viewpoint. Science educators are not yet certain whether, in such circumstances, they should take a positive stance, or present the facts as neutrally as they are able. Certainly care should be taken that non-white pupils are at least not deterred from studying the sciences, or of searching for jobs in science/technology-related employment.

Having raised the wider issues, we shall concern ourselves here with the more pragmatic problem of specific content, always remembering that it is not content in isolation that may be contributing to the underachievement of non-white children. In chapter 5, the importance of the communication process was discussed. Ethnic minorities have special problems in this respect, so that careful use of language becomes of paramount importance in a subject so prone to the use of specialised language and jargon as science.

Much current science content simply disregards the personal experience of white children, let alone children of other ethnic origins. It is here that science content needs to be scrutinised at the school science department level, where local teachers are best placed to perceive potential problems. In truth, the preparation of an 'ethnic content' for a science curriculum is no more than an extension of the principle of individualisation that has been stressed so much in this book. A workshop on the presentation of race in teaching resources produced the following guidelines, which might be useful for teachers developing their own materials (Brandt, Turner and Turner, 1985):

- avoid line drawings that caricature – photographs of real people are preferable, but these must be up to date
- avoid line drawings that are obviously European figures coloured black – features and body shape must be authentic
- avoid statements that imply a hierarchical relationship between countries of the north and south
- show ethnic peoples in positive, high-status roles that involve making decisions
- present up to date information about life in other countries, showing that advances in science and technology are worldwide
- where the discoveries and lives of great scientists are included, show that science is an international pursuit to which people of all races make important contributions.

It is not difficult to discover how a multicultural stance can be taken on particular topics when this becomes necessary. Using figure 8.1 as a starting point, the need for food for energy and

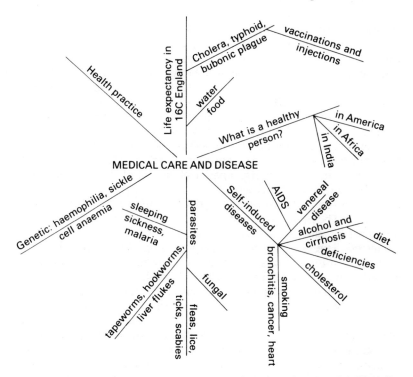

Figure 8.1 Some multicultural themes, exemplified through unit 1.8 (table 8.1), 'Medical care and disease'

growth (unit 1.3) can be exemplified by including content on the production of strains of rice to suit different climates, the dietary value of yams compared with potatoes, various dietary practices (Halal, vegetarian, Kosher), and the digestive disorders brought about by culture and diet. Figure 8.1 shows the modification of content for unit 1.8 (medical care and disease, see table 8.1) that might be appropriate in classes with a proportion of ethnic groups.

Gender and content

In chapter 6 it was noted that girls do not respond to the content of the science curriculum in the same way as do boys (Ormerod, 1975). Whilst both boys and girls enter school with similar interests in science, by the time they reach year 3 the girls' attitudes have deteriorated significantly more than the boys' (Reid and Tracey, 1985). This phenomenon has been widely reported, as has one of the hypothetical reasons thought to underlie it: girls struggling to establish their femininity reject what is perceived by them as a male stronghold (Kelly, 1979). Were this the only reason for the comparative underachievement in science by girls, then conditions that allow girls to study science in single-sex groups should result in some amelioration of the condition. However, this appears not to be the case (Harvey, 1985), which indicates that, at the very least, there are a number of cultural factors at work (Price and Talbot, 1984). In fact it is possible that there is, in addition to cultural factors, a genetic basis for this difference in attitude and achievement. Sex differences in a variety of intellectual abilities have been well documented, and in areas apparently related to science – for example, numerical ability, spatial ability, and problem solving (MacFarlane Smith, 1964).

Whilst attention to more 'girl-friendly' science content is important, it must be remembered that the content of the curriculum is but one part of the overall science experience to which girls are subjected. The GIST team (see chapter 6, pp. 104–5) introduced a number of interventions into schools in an effort to improve girls' attitudes to science (Kelly, Whyte and Smail, 1984). In particular, they instituted a programme of visits by women working in technical jobs; posters and worksheets about women's contribution to science; curriculum development to produce materials more oriented towards girls' interests; feedback on girls' participation in science lessons; and careers advice to girls. The impact on girls' subject choices was marginal – about 4 per cent more girls chose to study physics in the GIST cohort than in previous years – although the results of attitude tests were more encouraging. The point the authors make, however, is that schools alone cannot solve the problem of girls' underachievement in science and technology since 'they are enmeshed in the wider social structure. Parents, primary schools, peers and employers all influence children's attitudes to their choices. Teachers are reluctant to change their routines, and men as a group have no wish to relinquish their power'.

The sociological paradigm within which the research was undertaken does not, however, focus sufficiently on the role of curricular content in improving girls' attitudes, a point that is conceded by the researchers themselves. They perceived major anomalies in curricular content. For instance, they noted a 'tendency to approach topics from the point of view of a teacher who had an overview of the whole field rather than from the children's own interests', a warning we have consistently made in this book. They found one booklet on light for 12–13 year olds that started by discussing its wave nature! They were able to encourage the teacher to rewrite the booklet so that it began with eyes and moved on to more abstract topics, but were unable to persuade the teachers that major innovations were necessary. The introduction to electricity via the working of record players and cassettes was popular with girls, but never became established because the teachers felt it was 'not part of the normal approach to electricity'. In essence, they suggest four principles that should guide the selection of science content for girls:

- removing any masculine bias in the form of illustrations, language and examples; most science textbooks are biased in this way (Walford, 1983)
- linking experiments with other types of activity that girls enjoy, e.g. discussions and creative writing (see chapter 5)
- emphasising the application of science to everyday life before introducing difficult ideas, concepts and theories
- starting with topics familiar and interesting to girls, yet leading to an understanding of all science.

In the last case, for example, girls would be expected to respond more positively to the science of radioactivity if it were taught through its medical applications, where it is clearly seen as of benefit to society, rather than through abstract, theoretical content, or with emphasis on its application to heavy industry, such as power supplies and machines of war.

METHOD AND CONTENT

In line with the dual principles that less able children must take more rather than less responsibility for their own learning, and that content will always be influenced by other considerations (in this case appropriate method), we will introduce two techniques that have proved themselves as powerful learning devices with less able children.

Children as teachers

Jerome Bruner describes an experience common amongst teachers:

> Teaching is a superb way of learning. There is a beautiful story about a distinguished college teacher of Physics. He reports introducing an advanced class to the quantum theory 'I went through it once and looked up, only to find the class full of blank faces – they had obviously not understood. I went through it a second time and they still did not understand it. And so I went through it a third time, and that time I understood it.

It was in the last years of the eighteenth century in the United Kingdom that Bell and Lancaster first instituted their 'monitorial' system. So successful did the method become that, according to Bell in 1797, 'for months now it has not been found necessary to inflict a single punishment', and by 1816 there were at least 100,000 children in England and Wales being taught in this way (Allen, 1976). Whilst Bloom's taxonomy of educational objectives lay a century and a half into the future, Bell was recognising the gains that children involved in the method could make:

> By these means a few good boys ... teach their pupils to think rightly [COGNITIVE]..., and, by seeing that they treat one another kindly [SOCIAL], render their condition contented and happy [AFFECTIVE].

Gartner, Kohler and Riessman (1971) describe a particularly successful 'children as teachers' programme, developed in New York, and called Mobilisation for Youth (MFY). Not only did this experience catalyse a host of further programmes, in both America, Europe (Thelen, 1969), and the UK (Goodlad, 1979; FitzGibbon, 1978a), but it is particularly apposite since its concern in the first instance was for the underachieving child. At that time, in the early 1960s, 70 per cent of the 8 year olds in Manhattan's Lower East Side were reading below the average grade level. Amongst these children, the social distribution was also uneven, for it included 83 per cent of the Puerto Rican population, 77 per cent blacks, and 51 per cent whites. What is more, retardation of the reading age was both progressive and cumulative: by the time these children were 10 years old, 65 per cent, 34 per cent, and 14 per cent respectively of the races were reading more than three years below average.

The idea behind MFY was that local high school children would be sent to tutor other low achievers. There was considerable concern that the 13 year olds sent to teach the younger 8 year olds would suffer educationally in other areas of the curriculum, because of the time commitment of their teaching, and that the younger children would fare even worse when being taught by low-achieving 13 year olds than by their trained class teachers. So the gains and losses made by both groups of children were carefully monitored, in terms of both affective and cognitive changes across the curriculum, and involving the use of control groups in substantial numbers. As far as the reading skills of tutors and tutees were concerned, both showed gains after the five months of the experiment (see table 8.3).

The table shows that the control group – the children receiving no peer tuition – continued to lose ground: over the five-month period of the experiment, they made a gain of only 3.5 months in their reading grade. Those receiving four hours of tutoring per week, in contrast, began to catch up on themselves. That, however, is just

Table 8.3 *Reading gains made by tutees in MFY programme*

Amount of tutoring (hours per week)	Reading age of tutees (months below average grade level)		
	Before programme	*After programme*	*Gain in months*
4	−13	−7	+6
2	−13	−8	+5
0 (control)	−13	−9.5	+3.5

part of the story. The kind of reading gains made by the tutors, themselves underachievers, were remarkable by any standard. During the five months of the experiment, they made, on average, gains of 3 years and 4 months. These kinds of gains, particularly by the tutors, appear frequently in the literature (e.g. Fitz-Gibbon, 1977), and in many subject areas, including science and mathematics.

Just as remarkable as the cognitive gains are the changes in attitudes seen in these children. In another peer teaching programme, each tutor was required to keep a diary of experiences. Some of the entries make moving reading when it is remembered that these are tough, street-wise urchins from the slums of a large city (Portland, Oregon, in this case).

> Nov. 29. Sherry was stubborn, and wouldn't work today. We are working on division. She doesn't try anymore.
>
> Dec 5. I don't know what to do about Sherry. She still won't try.
>
> Dec 9. All of a sudden, Sherry understands division. She was pleasant and cooperative today. She was like she used to be. I guess I was wrong about her trying. She didn't understand and lost confidence. I am sorry I didn't understand the situation. She is a real good kid. I am never going to accuse her again.

Here we see that the tutee, Sherry, had not only learned to do something she had not previously been able to do, but that her whole attitude had changed. She regained her self-confidence, she 'was like she used to be'. The tutor recognises her role in this metamorphosis, and has gained valuable insight into her own self-perception. Of course, the reasons why peer tutoring should be such a powerful means of producing affective and cognitive gains, particularly in the tutors, is the subject of much conjecture. Indeed, many theoretical models have been used in explanation (Malamuth and Fitz-Gibbon, 1977). Suffice to say that the method does work, and has been used successfully in many countries, with many groups of children, and in many subject areas. But it is a particularly powerful device for use with less able children, and one that is recommended for trial in the teaching of science to the less able.

To conclude with a word of caution. The method is not a panacea for hard-pressed teachers! The tutors have to be trained, and their teaching programmes carefully devised. A degree of timetable flexibility is essential. Above all, extensive records evaluating the progress of both tutor and tutee have to be kept, which must not interfere with the degree of authority and responsibility given to the tutor. On the MFY programme, the tutors did the teaching in their own time after school, for which they were paid, in those days, the sum of $2 per hour.

Fitz-Gibbon (1978b) believes that tutoring on a one-to-one basis is the most beneficial for both partners. She suggests that there are a number of distinct organisational stages that need careful consideration when setting up such a programme:

1. Select the tutors. What are the characteristics to look for in a tutor? Can all children act as potential tutors?
2. Diagnose the tutors' needs. Are they weak readers; particularly poor at number work; poor observers; etc?
3. Select content to match each tutor's needs, e.g. *a narrative account* of the workings of a chemical plant; *interpreting data* and drawing charts, tables, etc; *observation of behaviour* patterns of gerbils, mice, the moon and sun, etc. Material can be negotiated according to the perceived needs (i.e. weaknesses) of the tutor.
4. Train tutors, individually or in groups, during lessons or at lunch times.
5. Locate tutees who can benefit from content. Often these will be younger, more able children.
6. Record the progress of both tutor and tutee, help tutor to devise more appropriate teaching devices, and provide feedback on how successful the programme is.

Group versus individual project work

The essential characteristic of project work is that it incorporates activities on the part of the children involved, the results of which are genuinely unknown before the work starts. Usually less able children feel happier working with a limited aim in mind. The theory, content, and organisation of individual project work is discussed in relation to an individual case study in chapter 9 (pp. 196–9). It must be admitted, however, that many underachievers are insufficiently motivated to be able to take responsibility for individual work of this kind, however short in duration. Under these circumstances it is sometimes possible to begin the process of motivation in science through group project work, where members of the group take responsibility for completion of certain parts of the project. If the project is negotiated with the children in the first instance, it is more likely to be successful. Good ideas can often come through local newspaper articles. For instance, a survey of burglaries over several weeks, pinpointed on a large-scale local map, might provide a number of hypotheses as to how they might be prevented. Measurements of density of housing, number and power of the streetlights, availability of roads for quick getaways, and frequency of police patrols in the various areas can all be correlated, each hypothesis being tested by a different group of children, until a comprehensive picture is built up.

One highly successful group project was performed at a national level throughout Britain in 1985. The pollution of rain water by acids from sulphur and nitrogen emissions is having disastrous effects on the lakes and forests of Sweden (Swedish Ministry of Agriculture, 1982), and there are signs of increased acidity in British lakes and rivers. The WATCH Trust for Environmental Education and the Field Studies Council of Britain organised a project called 'Acid Drops'. 'In back gardens all over Britain, thousands of children and their families braved icy conditions on winter mornings to collect their rainwater samples. Acid Drops ran for four weeks from mid-January to mid-February 1985 and its aim was to fill a real gap in scientific knowledge' (Paskell, 1985). The results were an outstanding success in gaining a picture of acid pollution in the UK. Not only that, but the children showed that valuable scientific information can be obtained by inexperienced people using low-technology kits. In September and October 1986, the project was repeated on a Europe-wide basis. This time the children sent in their results to the Forest Conservation Centre in Essex on BBC microcomputer discs. When the results have been analysed, they will be put on those same discs and returned to the schools, so that the individual school will be able to see its own contribution in the context of a genuinely international survey. The survey can then be followed up in schools with a more serious theoretical examination of pollution generally.

In years to come it is planned to investigate not only the effects of sulphur dioxide emissions from heavy industry, but also the effects of ozone from various types of nitrogen oxide (NO_x) emissions, mainly produced by the burning of fossil fuels, and radioactive pollution from power stations. The sampling will involve the susceptibility of living tissues to ozone (using a variety of tobacco plant) and radioactivity (using a variety of *Tradescantia*; see Ichikawa, 1981). Teachers can obtain information about previous, current and future project work from WATCH (see Appendix for address).

ORGANISING FOR CONTENT

Different learning experiences

As far as underachievers are concerned, we firmly believe that the success of the learning enterprise depends crucially on its organisation, and that better organisation can substantially raise the quality of their learning experiences and their ultimate levels of attainment.

The kind of curriculum we have proposed in this book does, of course, present the teacher with a wide range of organisational problems. If, for example, we really believe that children should be encouraged to 'construct their own reality' of content, as Barnes (1982) argues, then we must provide opportunities for small-group discussion work. If different children are to follow different learning routes and employ different learning methods, there will be complex organisational problems concerning record keeping. If children following different routes are to be provided with the necessary feedback on their progress, there will be further organisational problems. When all the constraints of the particular school environment are added (architecture, laboratory design, staffing, timetabling, and so on), the complexity of these organisational issues is immense.

Different children learn in different ways, and our departmental organisation must take account of it and must encompass a degree of flexibility to enable the teacher to modify and adapt to changing situations. Matching the learning experience to the individual learner could be seen by some as a form of streaming and, as such, may be regarded as discriminatory. We take the opposite view. We see individualisation as democratic. It is democratic because it seeks to ensure that all children, whatever their starting point, attain the target goals for science education. This process of matching and individualising, once accepted as legitimate, becomes a matter of careful organisation and management.

The curriculum planning model discussed in the introduction to part II comprises six essential stages:

1. Identification of the curriculum aims and objectives.
2. Appreciation of the characteristics of the learners and the establishment of an appropriate learning environment.
3. Selection of the teaching/learning methods appropriate to the learners and to the aims and objectives identified.

4. Selection of content appropriate to the learners and to the aims and objectives.
5. Organisation and provision of learning experiences to achieve the aims and objectives identified in (1) for the children characterised in (2).
6. Assessment and evaluation, which includes assessment of levels of pupil attainment, diagnosis of particular strengths, weaknesses and difficulties, and the evaluation of the worth-whileness and effectiveness of the learning experiences.

The organisation of particular learning experiences requires a degree of task analysis: What is to be learned? How is it to be achieved? What resources are required? How will success be defined and measured? In earlier chapters we discussed a number of principles underpinning good design – principles such as providing a degree of self-regulated learning, sequencing concepts from simple to complex, matching experiences to content and learners, ensuring that material is well structured, providing a variety of stimulus and activity, ensuring active participation, and providing rapid and effective feedback. All these matters must be kept in mind during the organisation stage. We suggest that the knowledge content of each module should be identified and coded, with a clear distinction being drawn between basic core provision and any extension provision. The second step in the design of learning experiences is the generation of possible learning activities focussing on that content, e.g. pupil experiments, teacher demonstrations, CAL packages, films, discussions, etc. It is recommended that the planning matrix (figures 7.6 and 7.7) is employed to ensure that learning activities are selected on the basis of their contributions to the attainment of attitudes and the fostering of heightened personal awareness (table 3.2, A1–A17) and the learning and experience of scientific processes (table 3.1, P1–P21), as well as for their contribution to knowledge acquisition. Thereafter it is a question of paying close attention to the issues of matching (method to content, method to child) and sequencing (logical development of concepts, good pedagogic dynamics, variety of stimulus, etc.).

In order to provide the kind of flexibility we require, it will be necessary to opt for resource-based individualised learning. For each unit within a module there will be additional sub-units, at a variety of levels of difficulty (some remedial, some extension) and designed to take account of particular interests and aptitudes. It is envisaged that the least able learners will spend most of their time on the basic units within the module, with occasional visits to remedial, enrichment, and special interest sub-units, whilst quicker learners will engage in a much wider range of sub-units. In order to match the learning experience to the particular individual, we need to know three things. First, we need to know the child's existing cognitive framework, and what process skills and attitudes he possesses. In other words, we need an up-to-date record in all three major areas of the science curriculum ('K', 'P' and 'A' components). Second, we need to know something about preferred learning style, linguistic skills, and general interests. Only when we are in possession of all these data can we be confident of designing the experience ideally suited to a particular learner. If we had classes of half a dozen children, such an idealised approach might be possible. In the real world, however, the best that can be hoped for is the provision of several routes through a topic. Skilful gathering of evaluative feedback during the learning process (i.e. individual formative evaluation) will enable us to modify the approach (simplifying, challenging, encouraging, changing learning groups, switching to another learning method, etc.) as the learning proceeds. This constant process of evaluation and decision making makes extraordinary demands on the teacher, but it is the only practicable way of optimising the learning experience for a particular child. A 'perfect' route cannot be planned out in advance for all children, and successful routing depends upon the use of sound organisational procedures.

The key to the provision of alternative learning routes through a module, and a degree of negotiated content, lies in the keeping of comprehensive records of the routes taken by individual children (see figures 8.3 and 9.6). With regular reviews of the record cards, teachers can identify and compensate for any deficiencies in experiences. It is the teacher's primary responsibility to ensure that a sufficient range of experiences is entered into, and that a sufficient amount of learning ensues. This role demands a high level of familiarity with the course, and high levels of understanding of children, together with a flair for organisation. In order to reduce the organisational complexity we propose that all children begin and end a module at the same time and, from time to time, with a common experience (a film, teacher demonstration, or whatever). Despite our strong advocacy of individualised learning, there is still value in group experiences. Individualisation does not necessarily mean working in isolation; indeed we are in favour of small-group work. It is the teacher's use of materials, exploitation of group dynamics, and subtle manipulation of teacher–child relationships that creates individual experiences and typifies the skilled professional teacher.

As the business of teaching becomes increasingly complex and demanding, as teachers become more accountable to others for their actions, so teachers have a right to expect increased support in the classroom. This can take many forms. The type of science education advocated here places new demands upon teacher trainers, science advisors and LEA support, often in the form of inservice training. Much of this is already happening, and the Department of Education and Science is spending large sums of money on inservice training in this field. It is unfortunate that there is, as yet, very little co-operative planning between teachers working in different schools, and it is hoped that the section in the Appendix on developments in science education (pp. 231–3) will make some small contribution to this end. The pooling of resources would be a spur to the implementation of resource-based learning schemes. There is much helpful advice in Beswick's (1972) book and in recent publications from the Association for Science Education (ASE, 1979b, 1985).

Records

The recording scheme we envisage has two major functions:

- to provide a record of curriculum experiences, and
- to provide a record of attainment, comprising knowledge (the 'K' component), process skills (the 'P' component), and attitude (the 'A' component).

The teacher needs to maintain a record card for each class that lists the experiences provided under the various sub-units within the modules, and does this in terms of the 'K', 'P', and 'A' components listed above. In a science department employing a substantial amount of individualised, resource-based learning it would be necessary to maintain individual records of the units attempted (figure 8.2), together with a listing of the experiences provided in each unit (perhaps as in table 8.2 above). In this way, the progress of each individual is easily accessed. Attainments of individuals in the three areas would be recorded on separate cards, and these would form the basis of a profile report.

As far as the 'K' component is concerned, a simple entry of mastery attained is sufficient (figure 8.2). Each topic should be represented at a number of difficulty levels – four is an adequate number. Some topics lend themselves better to higher-level work than others; the intrinsic nature of the content, children's and teachers' interests, available resources, and so on, can all be factors. Consequently, the number of difficulty levels will vary not only from topic to topic, but from school to school, and even from class to class. Some children will proceed only as far as level 1, others may go on to levels 2, 3 or 4. The basic minimal level of attainment, implied by the notion of scientific literacy, would be fixed in

NAME OF PUPIL	Module 1: Understanding ourselves — Unit															
	1.1		1.2		1.3		1.4		1.5		1.6		1.7		1.8	
	C	T	C	T	C	T	C	T	C	T	C	T	C	T	C	T
Aaron, Bev	*	*	*		*	*	*		*	*						
Briggs, Carl	*		*		*											
Caran, Del	*	*	*	*	*	*	*	*	*	*	*	*	*			
Donovan, Eam	*	*	*	*	*											
etc.																

C = unit completed
T = unit test has been satisfactory

Figure 8.2 Module 1 record card (units/tests completed), 'K' component only

advance. For example, if there are eight modules, four of them core and therefore compulsory, and four of them negotiable, we might be looking for achievement at level 1 in four of them, and perhaps level 2 in two of them. Ticks are made in the appropriate boxes of the individual record card (figure 8.3), and the mastery level is simply the total of scores on the unit tests. What is important is the principle of a minimum level of attainment for everyone.

As far as the 'P' component is concerned, there is much of value in White's (1979) proposal for a three-level grading system. The first level is 'achievement', where pupils demonstrate that they have acquired the basic knowledge and skills. At the higher level of 'mastery', knowledge and skills are integrated, enabling pupils to deal appropriately with novel situations. At the highest level of 'proficiency', the skills are so well practised that their performance becomes smooth and routine, rather like the proficient driver who can negotiate all manner of complex traffic problems without seeming to give the matter any great attention.

The 'A' component raises somewhat different issues. We feel that only positive items should be recorded, and that there should be some element of self-assessment. Developing self-awareness in a child and a capacity for self-criticism were identified as important curriculum goals (chapter 3). It would seem that self-assessment, in areas of cognitive and affective gains, encourages children's

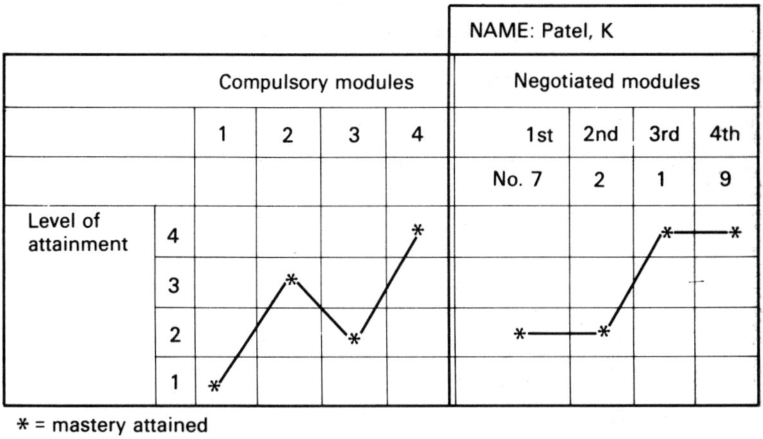

						NAME: Patel, K			
		Compulsory modules				Negotiated modules			
		1	2	3	4	1st	2nd	3rd	4th
						No. 7	2	1	9
Level of attainment	4				*			*	*
	3		*						
	2				*	*	*		
	1	*							

* = mastery attained

Figure 8.3 Pupil record card (details of tests), 'K' component only

awareness of their progress. In this way success is internalised and, in turn, encourages activities that will lead to even greater achievements. Rowntree (1977) and Sutton (1986) have valuable advice for teachers on the techniques of pupil self-evaluation.

Profiling

Pupil profiling is closely allied to graded tests: tests with short-term, precise objectives would be taken at regular intervals, thus assisting pupil motivation and bringing about a shift of emphasis towards goal-oriented and mastery learning. With the increased availability of microcomputers in schools, it is feasible to provide successful learners with instant accreditation. We would push this notion a little further and advocate the adoption of a test-when-ready principle, along the lines of the graded tests in music. This seems to be a natural consequence of the adoption of individualised learning strategies. A fully developed assessment scheme for a resource-based individualised science curriculum would comprise three phases: pre-assessment, diagnostic assessment and post-assessment (Hodson and Brewster, 1985).

Pre-assessment determines the appropriate starting point on the learning programme for the individual pupil; indeed, it may show that the pupil has sufficient skills to by-pass a particular unit and move directly to the next stage. Diagnostic assessment provides information on how well a pupil is developing, and serves to differentiate between those who are progressing well and those for whom intervention strategies might be appropriate. Post-assess-

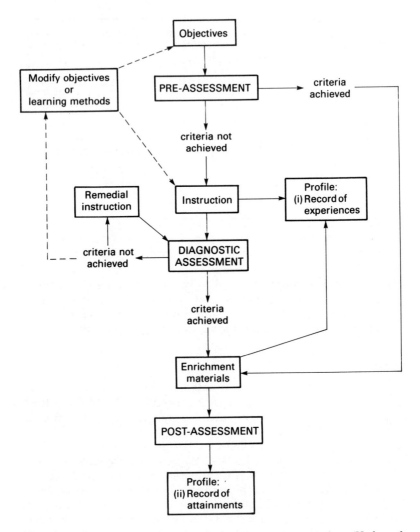

Figure 8.4 The relationship of assessment strategy to curriculum (Hodson & Brewster, 1985)

ment measures what the pupil has achieved in terms of intended outcomes, and it is this measure that forms the basis of the final science profile. The relationship of the assessment strategy to the whole curriculum is shown in figure 8.4.

The timing of the diagnostic assessment in the learning process is of crucial importance. There must be sufficient time both for the pupil to develop the skill or acquire the knowledge to be assessed, and for appropriate remedial activity if that should prove necessary. If this requires some reduction in content, then so be it. It is more satisfactory to have a smaller core of carefully selected material, in which children can have a high degree of competency, than to provide a rather larger amount of material that remains largely unmastered. The teaching of science remains very much an art, and the somewhat mechanistic approach to assessment and profiling that we are advocating is measuring only the more accessible gains that children make. At a deeper level, the sensitivity of the teacher should be allowed to modify the route that different children need to take through a science course.

Mastery learning

As discussed in chapter 7, the notion of mastery learning entails the establishment of an appropriate criterion score. It is necessary to know how well pupils must demonstrate a skill or display knowledge before it can be claimed that they have achieved mastery. Gronlund (1982) suggests the following checklist procedure for setting criterion scores:

- set mastery levels at 85 per cent correct
- increase level if essential for next stage of instruction
- increase level if essential for safety (e.g. mixing chemicals)
- increase level if test or sub-test is short
- decrease level if repetition is provided at next stage
- decrease level if task has low relevance
- decrease level if tasks are extremely difficult
- adjust level up or down as teaching experience dictates.

Whilst the pass level for mastery is arbitrarily set, it does involve the judgement of the teacher and feedback from trials. The acceptable standard is, to an extent, socially determined and is likely to correspond to rather vague notions of 'competence performance' held by those who have experience of the kinds of achievement that are possible by any particular group of underachievers. Certainly care must be taken to see that it is not too low, when learning is trivialised, or too high, where failure becomes the norm.

Extended assessment provision in favour of more profiling has certain advantageous curricular implications. It has long been argued that teachers organise curricular experiences primarily to meet the demands of the examination system; with the advent of the GCSE – an examination devised for all children including the least able – there is a real danger that the advantages implicit in profiling will be prematurely discarded in favour of a more traditional norm-referenced approach to assessment.

The introduction of profiling could be used as a vehicle for changing many fossilised curriculum activities. Certain teaching styles lend themselves to the assessment of process skills, so that the introduction of profiling in these areas would force a change in teaching style. By officially recognising and recording certain skills and attitudes, these attributes would be revalued in the curriculum and given greater consideration in lesson planning. In addition, teachers would gain new insights into their pupils. Good assessment and good profiling and record keeping would assist diagnosis of learning difficulties, prevent stereotyping and the rejection of 'non-academic' children, and provide a sound basis for counselling and guidance. The adoption of short-term goals, through a modular course structure, and the introduction of an incremental credit scheme lead to enormously enhanced motivation (see chapter 9). This comes about as much through increased self-awareness of strengths and weaknesses, improved staff–pupil relationships, and shared responsibilities for learning, as it does from curricular content *per se*. This is why we are discussing both content and organisation in the same chapter, for we are convinced that too much concern for content detracts teachers' energies away from other important areas.

Some anxieties

Hodson (1985b) has described some of the anxieties felt by teachers faced with the demands of profiling. Those teachers who are suspicious of the value of profiling claim that it is a matter of 'the assessment tail wagging the curriculum dog'. The counter-claim is that that is precisely what the public examination system has been doing for decades in the UK, and this is reflected in the concern over the university influence on public examinations at 16+, which has been particularly destructive for the group of children that form the subject matter of this book. Some teachers are concerned about the massive inroads that profiling will make into what they already see as inadequate science curriculum time. We would argue that that time is inefficiently spent at the moment, and the phenomenon of gross underachievement by some children in science is the only tragic evidence we need to supply in support of change.

The reluctance of many teachers to make judgements in non-cognitive areas for inclusion in publicly available profiles is interesting, because those same teachers frequently make those same judgements in confidential references. Broadfoot (1979) points to a major mismatch between teachers' expressed views and their actual practice: they profess to concentrate on cognitive criteria, yet over 50 per cent of the variance in class placement is, in practice, based on non-cognitive assessment. Moreover, these assessment data are unsystematically and often unconsciously gathered. We are aiming at a significant shift towards systematically gathered assessment data in these areas, and the compilation of profiles that fully represent each pupil's experiences and achievements across the whole curriculum spectrum, and experiences that are discussed with and agreed by the children themselves. In short, we are looking for an assessment scheme that matches the curriculum.

9

Case studies in three schools

The *Science at Age 15* report written by the Assessment of Performance Unit for science teachers (APU, 1985) makes the point that, in coming to a definition of 'science for all', it may not be appropriate to assess children just in terms of their scientific knowledge or their facility at manipulating scientific concepts. The definition may, it says, 'have to include an understanding of how pupils may use their scientific knowledge to investigate the world'. Such a mark of a scientific education seems to have become generally accepted in many schools that have a serious interest in the teaching of underachievers. This chapter takes as exemplars three schools that are using a variety of techniques in their teaching of science to underachieving children ('Castle Hill', 'Shey High', and 'Sidney Hall'). They have not been selected because they exemplify any particular teaching method or technique, although it may well be salutary that much of the philosophy underlying their techniques appears to be based on coming to terms, first and foremost, with the affective condition of the children rather than with their cognitive condition. All have a different notion of what is the 'less able' child. All depend heavily upon the professionalism of their science teachers. But most of all they depend upon the individual responsibility that the teachers are prepared to take upon themselves in the creation of an imaginative, exciting but still essentially 'scientific' science curriculum.

All three schools are LEA maintained – one in Lancashire, one in Manchester, and one in Salford. Two are large, 11–16, co-educational, inner-city comprehensives, and the third is a much smaller special school for children with moderate learning difficulties. Both of the comprehensive schools approach the teaching of science to underachievers through the 'alternative curriculum' strategy (ACS), a strategy with its origins in the EEC that first appeared in schools in the UK in the latter part of 1983. So successful has the strategy become that plans are afoot to involve not only the less able but, commencing in 1987, the medium-ability groups as well.

In essence, ACS involves the restructuring of the science curriculum into a number of modules, an organisational feature discussed in the last chapter (pp. 168–71). ACS is designed with

- less emphasis on traditional science subjects (biology, chemistry, physics, and even general or integrated science)
- a more conscious link with the adult world (this is specially well illustrated by some of the techniques and activities described for 'Castle Hill')
- emphasis on performance criteria as a basis of pupil assessment by the accumulation of credits, one at a time (see chapters 7 and 8); this is also the thinking underlying part of the GCSE science assessment procedures, and developments such as the Oxford Certificate of Educational Attainment
- a greater emphasis on co-operation in learning ('Sidney Hall' portrays this aspect particularly well)
- greater pupil involvement in the determination of their own learning, often by negotiation with teachers and sometimes with parents too.

One of the features of ACS is that it is centrally funded via the LEAs. Schools were given the option to bid to their LEA for a share of this money and, if successful, they were funded initially over a three-year period (1983–1986). The sums of money involved are not inconsiderable, but even so not all of the schools that made a bid were ultimately successful. Two of the schools discussed here made such bids – one was successful, and the other not. Both are currently running an alternative curriculum – one is richly endowed, the other receives no financial aid and has to 'rob Peter' (in the form of other subject departmental budgets) in order to 'pay Paul'.

For each of the three schools, the general background of science provision for their less able children is discussed. This is followed by a detailed examination in each school of two special features of science provision that seem particularly apposite to the teaching of underachievers, and that illustrate the principles of good curriculum design that we have outlined in earlier chapters.

SHEY HIGH SCHOOL

Successful in its bid for ACS funding, in 1983 Shey High School had 1,134 children on its roll between the ages of 11 and 16. There were 68 full-time-equivalent teachers. In 1982, approximately one-third of the pupils in year 5 left school with no examination qualification, or with CSE at grade 4 or below. Very few were able to find employment in what was, by then, the post-industrial society.

In September 1983, one-third of the new year 4 group (76 pupils) were offered the opportunity of working on the 'new' curriculum. Staff recommended pupils for ACS on the basis of one or more of the following criteria:

- they were perceived as bright underachievers
- they had special educational needs
- they were low achievers because of poor attendance
- they had difficulties in social adjustment.

The parents of potential ACS children were contacted, and a parents' meeting arranged. For a child to be accepted onto ACS, there had to be explicit agreement between teacher, parent, and pupil. A new style of curriculum was explained, the unique features of which will not appear remarkable to those who have read the earlier chapters in this book. These features included the following:

- a minimum staff–pupil ratio of 1:8, rising to 1:16, with a major emphasis on the development of more open relationships between pupils and teachers
- a tutor system that allows group tutors consistent access to their groups throughout the year. Half of each week is spent with the tutor, as well as three residential visits each year. This is more in accord with primary school practice than traditional secondary school practice, and gives time for the development of deeper relationships between teacher and taught
- a concentration on project work (constituting about half of the total ACS time)
- an emphasis on negotiation in the learning process. Not only is content negotiable, but so also is the method of assessment. Particularly useful were found to be pupils' self-assessment of their successes. External 'letters of credit' from the Northern Examination Association (NEA) take the place of CSE awards.
- a modular approach to science teaching is based on the *Science at Work* series published by Addison-Wesley (Taylor, 1979). However, it is important to realise that 'science teaching' is almost a misnomer in this respect. What has come to be seen as traditional science teaching has been replaced with much more of a cross-curriculum effort. Time ostensibly reserved for mathematics, project work, skills courses, and so on, may be used as 'science' time (although it may not). Thus it is not really fair to extract the 'science' component of the ACS *per se*, since it forms an integral part of the entire ACS package.

The success of the school in obtaining funding for its ACS proposal led to the appointment of two new full-time members of the teaching staff and associated clerical help. In addition, some £25,000 was made available for capital expenditure during the first year of the new curriculum. A breakdown of how this money was spent gives an indication of its cross-curricular influence, and also shows the considerable influence it had indirectly upon the facilities

Table 9.1 *Initial expenditure of capitation across the curriculum (Shey High School)*

Building	£ 761
Music	£1,625
* Computers and calculators	£1,240
* Outdoor pursuits	£1,000
O.H.P.	£ 258
* Photography	£1,395
Safe	£ 300
Typewriters	£ 800
* Video, TV and radio	£2,616
* Minibus	£3,000
Furniture	£5,471
* Tools and workbench	£ 270
* Canoes and trailer	£1,000
* Books, etc.	£2,449
* Residentials	£1,400
* Transport	£1,004
* Admin.	£1,306

available for the teaching of science in the school. Financial provision was disproportionately high in the first year to allow for 'front-end loading'. £2,600 of the total sum was spent as cash through an imprest account to pay for equipment and services that could not wait to go through the more cumbersome accounts procedures (table 9.1). Items marked with an asterisk are directly or indirectly available to the science department, which clearly did well out of the transactions. This kind of expenditure will be compared with that in school 2, attempting the same kind of innovations without funding.

The science staff involved in the ACS were volunteers. They joined a team of dedicated teachers whose overview of the educative process was not limited by a sole concern for the teaching of science. They could rely on these other team members for support in the total remotivation of the children involved in the scheme. But it remained the responsibility of the individual science teacher to perceive where in the scheme of things advantages to the teaching of science could accrue. Using the video equipment to show a film of 'Jaws' might form the starting point for a group project on sharks. A trip to the Welsh coast, fishing for sharks in the form of dogfish, would involve the use of the minibus, and could indeed form part of a residential weekend. Such unusual ventures formed part of the negotiated content. Pupils would 'pay' for such experiences when they had to knuckle down to more laboratory-based courses such as 'You and Your Mind'. But, even here, the children were expected to face the real world. In work on the senses, in particular touch, they were expected to design and to make a set of dominoes for use by blind children. Thus, the choice of materials and skills development were retained as viable instructional objectives. Remarkable feats of imagination were involved in this particular exercise, resulting in the production of sets of dominoes made from wood, ceramics, metal, and even edible biscuits with currants as spots. The winner ate all!

There are two particularly interesting features of the Shey High School ACS.

Project work

The aim of project work in science is above all to enable children to follow up their interests in specific topic areas. This involves them in taking responsibility for their actions, and in taking the initiative in order to get matters accomplished. An alternative form of group project work was discussed in the last chapter. Individual project work is an almost ideal technique for encouraging a number of affective gains in children, such affective gains as increased self-perception, persistence in the face of temporary set-backs, and the confidence to be found at the completion of a piece of work perceived by peers, parents, teachers and self alike as having intrinsic merit. It is the development of such personal qualities that goes such a long way to improving the quality of lifestyle that more able children, and most science teachers, are able to take for granted. An attempt to show the relationship between some of the stages involved in project work, and their associated affective gains is shown in figure 9.1. Skills such as letter writing, the use of scientific equipment, reading, and discussion are also encouraged. The technique is much less effective in the development of cognitive gains.

Experience at Shey has shown that, for the less able, projects of about half a term in length are ideal. This is because the end is in sight almost from the start, although one teacher explained that 'once the pupils become well motivated, much more was involved in the projects than appeared at first sight. Many could have extended well into a second term'.

The first principle underlying successful project work is that, once a child has selected an area for study, the teacher feels happy that a positive outcome can be achieved. The study must lie within the competence of the child, or the small group of children involved, if possible allowing early indications of success to encourage further work. Negotiation between teacher and pupil is therefore essential; there is no such thing as 'compulsory' project work in science. Even when the problem as been identified, the children's tasks will still require a modicum of structuring. Seminar work is particularly helpful in this respect. A certain part of each week given over to discussion among the children on what they have accomplished over the last week, what their intentions are for the next week, and any problems they envisage can be remarkably successful, provided all the children realise that they in turn will have to do some talking. Children become avid listeners when they understand that not only are their opinions sought, but there is a good chance that they will be acted upon.

Figure 9.1 indicates the major decisions made by a group of two children who decided to investigate the distribution of fruits from a sycamore tree during a six-week period in October and November. Figure 9.2 summarises how they went about the field study part of the project.

There are many forms a project might take. This one was successful for a number of reasons:

- *Organisation.* Not a great deal of specialised equipment is required, or is at risk. Each child was 'given' a small sum of money for spending on the development of apparatus and other materials. In this case a string transect and two 1 foot square quadrats.
- *Limited topic.* The distribution of sycamore fruit is completed within a six-week period, so that the data-collection period is limited. Graphs of the distribution can be built up on a weekly basis, so that the children can begin to see patterns developing fairly quickly.
- *Open nature of topic.* There is no 'right' or 'wrong' answer. If the distribution differs from week to week, then it differs from week to week, and there must be a scientific explanation for the phenomenon.
- *Flexible extensions are possible.* The basic question can be as simple as 'what is the furthest distance a fruit travels from the parent tree?' At a stage further, questions can be asked about the numbers distributed at different distances from the tree, and in different directions. Are the fruits furthest from the tree different in any way from those nearest the tree? This could involve measuring the surface area of the wings, and even perhaps considering the mass:surface area ratios (in this case, it did so). Weather data can be obtained from the local meteorological office, or indeed collected by the children. The number of additional measurements that can be made are limited as much by the imagination of the teacher and the children as by the inherent nature of the topic material.
- *Ease of access.* Even inner-city schools have access to parks and streets where such trees exist.

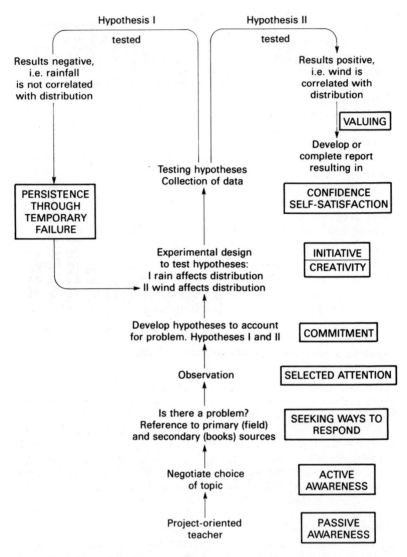

Figure 9.1 Relationships between affective gains and stages in project work

- *Extension work in the laboratory.* In a successful effort to simulate natural distribution, a different group of children decided to reverse the direction of a vacuum cleaner, and blow the fruits down the school steps.

Biological, technological, and social science subjects provide rich sources of materials for project work. At Shey, one social science

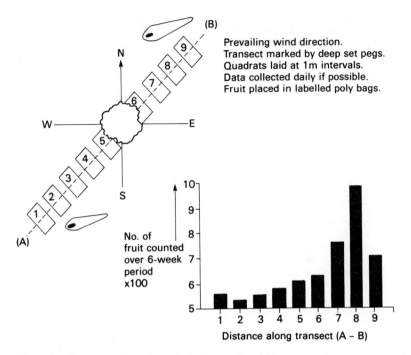

Figure 9.2 Summary of sycamore fruit dispersal project

project took the form of an 'age awareness' topic. Children surveyed the actual arrival of public transport at two hospitals, and compared it with the published timetabled arrivals. Elderly people from the out-patients department were asked to complete a short questionnaire on their opinions of the transport facilities available. This project not only involved the writing and analysis of a questionnaire, but allowed for the development of communication skills, and helped sensitise children to the problems of the elderly.

Residential experiences

Each child attended three residential courses in the first year of ACS. The science staff involved took an active interest in the content of these courses, but once again it is important to see the science as part only of a broader spectrum, whose aims and objectives were almost solely (at least in these early stages) concerned with the development of affective and social gains by the children. The use of canoes built by the children themselves in CDT lessons, a visit to a glass factory, and a talk on acid rain formed the science and technology component of one weekend. On another, the children were able to take photographs of the night sky in order to illustrate teaching sessions on stars and planets. The influence of man on the area (the English Lake District) was exemplified by an examination of pollution markers in a stream, copper mine spillage, and woodland management. Not all the children took part in all of these experiences, so that displays of collected materials and charts were able to form the basis of short talks by the children. All the children were introduced to one new experience, be it canoeing, abseiling or tenting out. They negotiated their involvement in the various activities, but they also negotiated their share of the work, which included preparing and cooking meals and washing up. One science teacher at Shey reported of her experiences after a year of residentials:

> Away from the classroom, everyone has been much more relaxed and everyone has had the chance to shine. Different members of the group have been the leader. Those who do not enjoy any form of academic work have excelled in practical situations. The group have become more aware of each other's strengths and weaknesses and are more considerate and helpful. As the confidence of the individuals has grown, the organisation of these trips is being left more and more to the pupils, as they now know what needs to be done and rarely report to the tutor what has been decided. The concept of 'negotiation' is now understood. Pupils are able to negotiate, for example, the amount of free time they should have, where they should be allowed to go in the evening, or if they should be allowed to smoke. The effect of this has been beneficial in the classroom and the idea of 'our group/our room is best' has produced a competitive spirit in all four groups. However, it must be accepted that taking a group of twenty inner-city children away from their environment is not as easy as it may seem. Problems are bound to arise, and indeed we have had some very unpleasant experiences to deal with. The most serious of these have involved theft and suitable punishment for those involved. Less serious incidents have included, for example, an individual who would not help with his share of the work, and one who refused to take part in an activity. This causes bad feeling with the group and produces an uncomfortable atmosphere for the others to work in. But the residentials were full of 'magic moments' – I look forward to more of the same next year.

The success of residential weekends such as those described above depends upon a close attention to the details of organisation, and for this purpose a checklist of procedures is given (figure 9.3). Above all, the science teacher must know what facilities are available at the chosen site, and a prior visit to reconnoitre potential usage is important. Finally, teachers should not be discouraged from attempting such residential stays by any apparent lack of

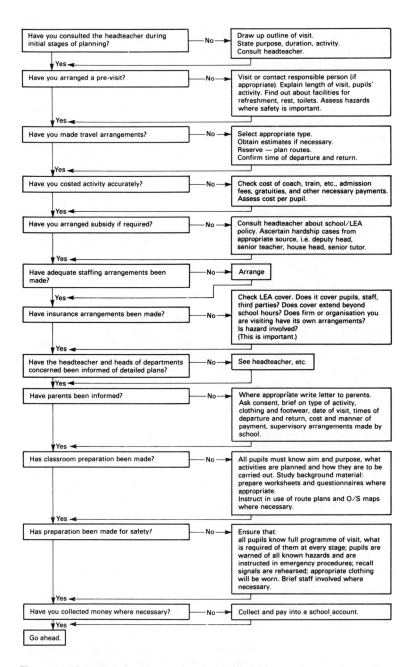

Figure 9.3 Flow chart for organisation of residential course

interest on the part of the children themselves. Older children have often become conditioned to dislike schooling in all of its aspects. On one field course, organised by one of the authors for socially maladjusted children, comments such as 'boring' were heard consistently during the three days the course lasted. On return to school however, letters were received from children who had been unable to attend, asking if they could have priority on the next occasion. Contact with the teacher revealed that those who had attended the course had been prone to exaggerate the length of the leeches found in the lake and the fun to be obtained from collecting moth traps at 4 o'clock in the morning! It is worth stressing at this point that more rigid guidelines for the general conduct of such courses are being laid down as a result of the death of some children at Lands End in 1985, and a recent court case that led to a teacher being fined. Teachers planning to take residential courses should check carefully that they fulfill all the requirements laid down by their LEA.

CASTLE HILL HIGH SCHOOL

This school was unsuccessful in its bid for ACS funding, but the staff had become so convinced of its suitability for their children that they decided to proceed regardless. Castle Hill has 937 pupils between the ages of 11 and 16. No extra staff were available to man the innovation, so that producing a viable pupil:staff ratio for ACS had serious ramifications for the ratio in the rest of the school. For this reason the numbers selected for ACS were comparatively low – about 30 from a year group of 180. The staff:pupil ratio for ACS pupils was thus kept to 1:7 in the first year, although it has since increased to 1:10 as the numbers increased to 60 in the second year.

Castle Hill is typical of many inner-city schools in that its IQ distribution shows a marked skew to the left (in this case eleven points below average, at 89 points), and this gives some indication of the IQ levels of the 30 or so children who elected to go ACS. Originally, 19 of the 64 staff requested to be involved in the experiment, and these teachers spent some 15 months in preparing the course. The head is particularly proud of the fact that this group of teachers met regularly on a fortnightly basis over the 15-month period, and there was never an occasion on which less than 15 of the 19 staff were in attendance. A mere £3,000 was allocated to the project, all of which had to be earmarked from the normal school budget.

In principle, the organisation is similar to that in Shey High School, and the curriculum was specifically designed to

- emphasise the practical application of science knowledge and skills
- relate learning to the real world
- facilitate active pupil participation and increased motivation by constructing short units, relying on negotiation between teacher and pupil in the development of pupil profiles
- reduce rigid subject divisions
- include residential experiences
- promote a positive self-concept in pupils
- be supported by regular counselling and personal assessment.

This last point is seen to form the key to so much of what was aimed for in terms of the personal development of the children. The head explained:

> Counselling sessions provide pupils with a stable point of reference and a continuity of support where advice, guidance and encouragement are essential elements. Problems can be discussed, self-evaluation encouraged, targets set and reviewed, and feelings of self respect re-inforced by the demonstration of worthwhile tasks having been accomplished.

The selection of the 30 pupils was based on their known negative response to the normal school curriculum. They are characterised by their short attention span, tendency to truancy and their behavioural problems. In spite of this, many of them had a flair for working with their hands. The science component of the ACS is inter-disciplinary with a tendency to be technological in nature. As with the other schools discussed in this chapter, two of the highlights of the science contribution are considered in more detail.

Real-world activities

One of the most important criteria embedded in the Castle Hill ACS philosophy appears second in the list above as 'relate learning to the real world'. In a genuine attempt to satisfy this criterion, many of the 'science' units appear technologically oriented. Children opting for the 'Energy in the Home and Society' unit, for example, still have to cope with such traditional-sounding science as 'electrical resistance and the production of heat energy', and 'electricity generation in power stations using various power sources'. The difference between this unit and traditional physics, however, is that only theoretical information that is required for the 'manufacture of a solar panel for use as a water heater' is justified. In the end, the success or failure of the pupils is measured by the success or failure of their solar panels in the heating of water in a

CASTLE HILL HIGH SCHOOL

NAME _____

MODULE ____ ENERGY IN THE HOME AND SOCIETY _____

UNIT _____

	COMPLETED	STAFF SIGN
Can use a water-bath correctly		
Can set up apparatus to instructions		
Can record observed qualitative results		
Can wire an electrical circuit to given diagram		
Can read electrical instruments accurately		
Can tabulate results correctly		
Has made an electrical water heater		
Has produced a flow diagram for a power station		
Has produced a poster about nuclear energy		
Has produced a design for a solar panel		
Has made and tested a solar panel to meet the design		
Teacher's comments on displayed work		

STAFF SIGN: _____

PUPIL SIGN: _____

DATE: _____

Figure 9.4 Profile for 'Energy in the Home' unit

water bath. On the road to this tangible end-product, the children will have demonstrated such skills as the 'wiring of a circuit to a given diagram'. Figure 9.4 shows the profile that the pupils will eventually include in a bound booklet as evidence of knowledge and skills.

Other science units incorporate opportunities for pupils to display their creative talents. The 'photography' unit, for example, requires not only that children learn to use a camera, light meter and flashgun, but that they understand some of the elements of perspective and composition. As far as science is concerned, they learn about the effect of light upon chemicals, the focussing action of convex lenses, and how to calculate aperture against shutter speed.

Some children learn their approach to science through model making. The child who built the model of the water mill (figure 9.5) was motivated in the first instance by the opportunity and encouragement to build the model. Negotiation had, however, provided the impetus for learning about the gear systems in the model, and how they worked in a real mill.

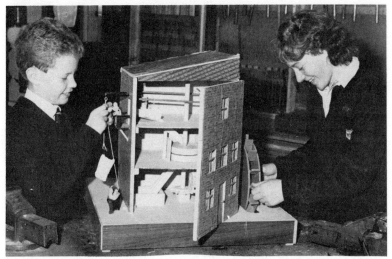

Figure 9.5 Model of a water mill made by Kevin at Castle Hill School (photograph by David Griffiths, University of Manchester)

The 'motor mechanics' units are reminiscent of the kind of curricular content commonly seen 30 years ago in secondary modern schools. They involve, for example, the stripping down of a clutch, and later an entire gear box. But, as before, success is measured by the fact that the child is able to drive the car around the school field after it has been rebuilt, and it is this activity that constitutes the 'real world' test. Again, during the process of replacing the clutch, the child learns to read plans, discover how a clutch operates, diagnose simple faults, remove and replace master and slave cylinders, and so on, all of which is recorded on the profile of achievement, signed by the teacher, then counter-signed by the pupil, and ultimately by the external examiner.

Small businesses

In the real world, men and women work in order to make money. A unique feature of the Castle Hill ACS is the motivating effect it has on children who are given the opportunity to set up their own small businesses and to keep a negotiated proportion of the profits for themselves. In the process of running their chosen business, the children have to give attention to a large number of factors. First they have to test the market to discover whether or not their proposed product is saleable. They then have to cost the purchase of their raw materials, machine tool any apparatus required for the production line, work out labour costs and advertising costs, and make decisions about quality control. If they take on a poor labour force, or 'employ' too many of their friends, they will lose their profits in excess wages. The small businesses are an exercise in production and profit. Successful businesses at Castle Hill have been in printing (selling programmes at school functions), cookery (a four-course meal to the governors), advertising (after negotiations with local businesses), Christmas decorations, market gardening, and the making of badges.

The children are keen to 'take advice' on the most cost-effective materials to use for making badges, and it hardly seems as if they are being taught about the science of materials. The badges did not sell at first because they were too expensive, and the other children in the school could not afford them, but eventually a compromise was found that enabled a small profit to be made. The cookery unit required the selection of an appropriate menu based on an understanding of calorific values and other components of a balanced diet. Safety and hygiene were taught as part of the preparations for the waiter and waitress service, and the washing up and cleaning of apparatus afterwards. Costing involved a great deal more than the presentation of the final bill to the customers! The market gardening small business was fortunate in being able to make use of the school's splendid heated greenhouse. During the production of *Geranium*, spider plant and *Kalanchoë* cuttings, the children learned about the water relationships of plants (*Geraniums* will not survive under the same conditions as *Kalanchoë*). They discovered the effect of disease, especially damping off fungus, which led to a study of fungal infections generally. They also read about fertilisers, and how different minerals had different effects upon plants. Geraniums coming into flower did not require copious (and expensive) doses of liquid nitrogen for example. Mustafa, a demotivated and difficult 14 year old at the beginning of the ACS experiment, has become a highly competent manager and gardener, to such an extent that at school open days the city Parks Department is no longer required to supply the school hall with a fine display of flowering potted plants. Sales are booming, and Mustafa has become quite philosophical about the vandalism that has, on several occasions, prevented him from expanding his business through an allotment located in the school grounds.

SIDNEY HALL

Sidney Hall special school caters for 176 children with moderate learning difficulties between the ages of 2 and 19. There are 22 full-time teachers at the school, and six nursery nurses, none of whom is a trained science specialist. About 90 of the children are between the ages of 11 and 16, and for this age range the staff:pupil ratio is 1:10. All the children at the school have been referred by educational psychologists and other professionals. As a result of the 1981 Education Act, all the children are there as a result of explicit agreements made between the LEA and the parents. Generally speaking, then, the children at this school have been more carefully screened than those selected for the ACS at Shey and Castle Hill. They are a more homogeneous group, although the science expertise amongst the staff is low. About half of the children show evidence of emotional maladjustment. Indeed, of the 176 children in the school, only about 30 show symptoms of slow learning that are not associated with some other handicap. Much of the curriculum is organised around the need to build up self-confidence in the children, many of whom come from culturally deprived backgrounds.

The children often lack some of the most basic experiences to be found in the natural world. At a residential week by the sea, the head was surprised to find how few of them realised that the sea was tidal. Some children were frightened when they thought that the sea had 'disappeared'. It is on such courses that the children discover that attention to personal hygiene can be a pleasurable experience; the school now supplies its own large soft towels, for instance, having discovered that some children did not like washing because of the small, hard and dirty towels they had brought away with them. Even such experiences as basic as seeing food had been denied to some. Seeing a piece of beef ready for cooking one day, a child assumed it was a turkey, since that is what they had eaten the evening before. The teaching of science to such experientially deprived children can take nothing for granted. Work on nutrition for example, a concrete experience for most children, becomes a highly abstract topic when the diet of the children is mainly hamburgers, fish fingers, and chips.

Being creative and keeping records

The science taught in this school is often immediate. It depends much upon the ability of teachers to recognise when a child's interest has been aroused by some natural phenomenon, however commonplace it might seem. Thus, the syllabus cannot be strictly prescribed by agreed content. It is essential, then, that teachers are aware of the different kinds of skills and processes that are being taught, and that they become adept at recognising these, for it is these that are ultimately more important that the content *per se*. The only way in which this can be effectively accomplished is by the keeping of accurate records, as we have argued in chapter 8.

Figure 9.6 shows how Ian's science education progressed over a five-week period in the early months of 1987. A glance at the horizontal rows shows that he was not involved in any experimenting or interpretation of data during this period, and this provides a reminder to the teacher that these two skills need to have particular attention paid to them over the next few months. He did, however, get involved in some worthwhile observation, measuring and communicating, and there is some evidence that he is beginning to be able to predict the outcome of experimental work. It should be stressed that the pattern of the teacher's recording on Ian's record sheet is fairly dense when compared with the patterns of some of the other children in the same group!

The kind of content that produced these results on Ian's record card is described below. There is no implication that the same content will be taught next year. By then the teacher will probably have discovered some other experiences more apposite to a different group of children.

On a cold day in winter, in this Lancashire town high up in the Pennines, the children were taken for a walk in the local park. There they looked at snowflakes through a hand lens and saw a thick layer of ice on the pond. The previous day, the teacher had made some ice balls by filling rubber balloons with water under pressure, sealing the balloons, and leaving them in the deep freeze over night. The rubber was then cut away to produce solid balls of ice. The children were encouraged to feel an ice ball, and to get their hands cold. Later, as it began to melt, their hands got wet. They were experiencing 'change of state'. Would the ice balls melt faster when put in warm rather than cold water? What would happen to the ice when it melted? They decided to measure the height of the water in a water bath before the ice ball was put in it, when the ball was first added, and after the ball had melted. A ball was dropped onto concrete and it smashed. Would the small pieces melt as rapidly as the large pieces? The way in which the ice floated in the water bath reminded one

Figure 9.6 (*opposite*) Ian's record sheet for January

SCIENCE RECORDING SHEET

TOPIC	Awareness of weather						Name: Ian		
DATE/PROCESS	6.1.87	9.1.87	13.1.87	16.1.87	20.1.87	21.1.87	23.1.87	27.1.87	28.1.87
OBSERVING	Visit park & look at snow c lens	Look at ice 'balloon'	Observe ice in water tank	Looking for signs of wind: smoke	Today's weather	Feel cold, warm, hot water	Fabrics for degree of waterproof	Steam from kettle – condensation	
MEASURING	Weighing ice	Weighing melted ice	Time for ice to melt in mins./sec	Wind direction using compass	Using wind meter	Using thermometer in classroom	1 teaspoon of water		Using digital thermometer
COMPARING		Weight of ice before & after melt	Depth of water before & after	Strength of wind (verbally)	Early a.m. with p.m.	Feel of warm/hot water	Efficacy of waterproof	Amount of water before/after boiling	Temperatures
INVESTIGATING		Properties of ice		Handling feathers in wind	Testing wind meter outdoors	Temps in classroom	5 fabrics for waterproof-ness		Temps outside classroom
CLASSIFYING				Wind forces	Today's weather for record		5 fabrics		
DESCRIBING	Language of winter	Types of ice - wet, dry, sharp	Changing nature of melting ice	Wind noises, forces etc.	Today's weather	How it feels	What happens to droplets?	Condensation & evaporation	
RECORDING AND COMMUNICATING	Short write-up of activities	Discussing properties of ice	Write up	Discuss & draw Beaufort scale	Fill in weather record	Worksheets	Draw droplet patterns	Worksheet	Discuss
CONSTRUCTING HYPOTHESES	Will ice weigh more melted?	What will happen when dropped?	Depth of water alters with melt?	Which way is the wind blowing?					Will temps differ?
PREDICTING					It will go fast in high wind	Hg will rise up thermo.	Which fabric makes best raincoat	Less water after boiling	
INTERPRETING DATA									
CONTROLLING VARIABLES			Frequency of recording Temp of bath		Wind meter kept at k height		Test with same amount of water		
EXPERIMENTING									
COMMENTS	Most rows contain at least some entries, which shows that Ian is gaining experience in a wide area of science. However, there are two rows, 'Interpreting data' and 'Experimenting', with no entries at all. Next month his programme must include opportunities for him to do some experiments, both in small groups and on his own.								

child of icebergs. Could they measure how much of the ice was above, and how much below the water surface? Would it make any difference if the ice were made of salt water? How could they find out? They filled some more balloons with water in which copious quantities of salt had first been dissolved.

One day the children came in from the playground after a heavy rainstorm; some of their clothes had let the water through, while others had not. This simple observation formed the basis of a two-week investigation, at the end of which a child had written:

> We wanted to find out which fabric made the best raincoat. We put each piece of material over a jam jar. Then we measured 2 teaspoons of water for each jar. We put the water on the material. We watched to see if the water dripped quickly through the wool. The plastic coated cotton was best.

Embedded in this simple piece of work it is possible to isolate a number of scientific skills and processes to which the children were exposed, although very little science 'content' as such has been learned:

- *Observation*. Some fabrics let in water, whilst others did not. Why?
- *Devising experiments*. What materials shall we test? How shall we test them? Note the simplicity of the apparatus – jam jars and teaspoons.
- *Measuring*. Using teaspoons. How much is collected in the jam jar? How do we measure this? Over how long a time period? Evaporation?
- *Controls*. Using the same amount of water, over a similar period of time. The teacher referred to this as a 'fair test'.
- *Interpreting data*. Drawing bar charts of the amounts of water getting through the fabric is an inverse measure of the waterproofness of the fabric.

The investigation was open-ended in that it gave opportunities for further work, which on this occasion was not followed up. The teacher could have gone on to develop the skills of hypothesis formation and prediction. Why are certain materials more waterproof than others? Is it related to the size of the pores? Using a microscope would have given the answer. Grading some untested materials by pore size could have led to predictions about their waterproof potential. These predictions could then have been tested by repetition of the jam jar experiment.

Not all of the science at Sidney Hall is as *ad hoc* as that described above. Serendipity cannot always come to the rescue. There is a syllabus, based on the Addison-Wesley series *A First Look* (1982), which

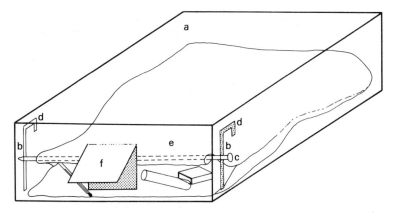

'a' Cardboard box, approximately 40cms x 50cms x 15cms

'b' Vertical slits are cut in both sides wide enough to house 'c'

'c' A large wooden knitting needle

'd' Slits shaped at top in such a way that 'c' can be kept at the top
without falling down

'e' A towelling bag, open at its front end, with its upper lip sewn
along its length around 'c' and its lower lip attached permanently
to the base of 'a'

'f' A flap, about 10cms square, cut into the front of the box

The bag contains a number (about eight) of items, e.g. a thermometer, a test
tube, a matchbox. The needle 'c' is lowered to the bottom of the slits, thus
closing the bag. When a hand is put through the flap it can only feel the objects
in the bag through the thickness of the towelling. Afterwards, the needle is
raised to the top of the slit, opening the bag. A hand put through the flap now
enters the bag, and can feel the objects directly.

Figure 9.7 The feely box

describes science suitable for use by 4–8 year olds. But it is a feature
of the science taught in the school that even work taken from
published material can be tempered by the creative skills of the
teachers. One teacher had devised a demonstration to show the
importance of texture in the sense of touch. A number of articles
were placed in a cardboard box (figure 9.7), which had been
modified to contain a piece of towelling that could be raised and
lowered over the articles without revealing them. Two teams were
formed, and the game was to score as many points as possible by
identifying such objects as a matchbox, a pencil, a wooden brick, a
conical flask, and so on. The towel was then lifted by raising the

large wooden knitting needle in the vertical slot, and the children again felt the articles. 'Before scores' and 'after scores' introduced an element of numeracy into the exercise.

Sometimes the children are better than the teachers in devising procedures. Why do some Smarties change colour as they are eaten? Lisa's idea for getting the colour off the Smartie and onto the filter paper for chromatography shows a capacity to cope successfully with more than one variable at a time!

> dip a straw into some water and so put a drop on top of the Smartie and then press the wet side onto the blotting paper and then eat the Smartie.

Pupils as teachers

It is a short step from the open approach to science teaching practised by the teachers at Sidney Hall, where so much reliance is placed upon the production of workable ideas from the children themselves, to a teaching technique that over the last two decades has received wide acclaim (see chapter 8, pp. 177–80).

One day an episode occurred that in an ordinary school might have served as an example of 'an experiment that went wrong'. The children were working in pairs. Some sodium bicarbonate had been placed in balloons, and the children were fitting the necks of the balloons over test tubes containing vinegar. As the bicarbonate powder fell into the test tubes carbon dioxide gas was emitted, and the balloons began to inflate. The teacher said 'they were delighted when one balloon flew off, scattering its contents all over'. After break, the children wanted to show the experiment to their friends, and a discussion started with the class teacher as to how they might do this. Were they sure of the amounts of powder to use that would cause the balloons to fly off? What would they say if their friends asked them how it worked? Does it happen with all household powders and liquids? The children decided that they did not know enough about the process to ensure either that the experiment would 'work' as they wanted it to, or that they could answer the multiplicity of questions their friends might ask.

In order to prepare for their teaching exercise, they developed a list of questions they thought their friends might ask, and devised further experiments with different powders. They realised that different combinations of liquids and powders might produce differing quantities of gas, and at different rates, and that the concept of a 'fair test' meant that equal quantities of powder would need to be added to equal amounts of liquid. They decided that the rate of gas production could be measured by recording the amount of time it took for the balloons to fly off the test tubes. In order to obtain equal quantities of powder and liquid they used lolly sticks as spatulas and teaspoons as measuring cylinders.

After practising with a number of powders and liquids, they eventually decided to use three combinations – baking powder and vinegar, baking powder and lemon juice, and yeast and sugar solution. They drew a chart on the blackboard, and filled in the time taken for the balloons to fly off the different test tubes. Each child demonstrated the three combinations to one friend. They took their responsibilities very seriously, and were concerned to make the experiment work properly. They were able to explain to their friends what was happening. The teacher said 'they executed the experiment and explained it without any help from me, in a logical, sensible manner. I was proud of them'.

With hindsight, this exercise can be criticised in the light of the criteria discussed in chapter 8, although it is abundantly clear that these less able children were quite capable of taking seriously the challenge to teach their peers, and derived pleasure and benefit from so doing. In this case it might have been better if the practical work had been followed up by the production of some written material, possibly on a word processor. Once again, it is not necessary that all the children produce individual work – all could have contributed something to the final printout. Follow-up work, relating the chemistry to various industrial and domestic processes – bread and wine (and vinegar) production, for example – would also have been relevant. Benefits to the tutors usually accrue over a lengthy spell of teaching, say half a term for two–four hours a week, and it is best to select specific areas of weakness in tutors, and have them teach in these areas. Above all, it is important that the class teacher monitors the gains that both tutor and tutee are making, in order both to modify the teaching syllabus if necessary, but also to provide accurate and encouraging feedback to both members of the relationship.

Current initiatives in context

THE GCSE CONTEXT

The publication of this book comes at a time when teachers in secondary schools in the UK are facing the challenge of their first year of teaching the GCSE syllabuses, and are exploring at first hand the implications of the changes that have been made. The GCSE has its roots in the Great Debate, inaugurated in 1976 by the then prime minister, James Callaghan. A year later, in her Green Paper (DES, 1977), Shirley Williams made plain the fact that government itself intended to take more central control of curriculum matters. In respect of the science curriculum, the government was particularly concerned that the nation's children should be prepared more adequately for a rapidly changing world of science and technology. This carried with it ramifications for not only the type of science syllabus that would best procure this end, but the quality and quantity of the scientific and technological know-how of the children who were to follow the courses.

The new single system of examinations at 16+ was announced by the Secretary of State for Education and Science, Sir Keith Joseph, in the House of Commons on 20 June 1984. The main features of the new examination system that pertain to less able and underachieving children are fourfold:

1. All syllabuses and assessment and grading procedures must be based on 'national criteria'. In effect, this implies a minimal common core science syllabus for children of secondary school age. For instance, one of the criteria for the science syllabus states: 'At least 15 per cent of the total marks are to be allocated to assessments relating to technological applications and the social, economic and environmental issues which should pervade all parts of the examination' (DES, 1985d). To this extent the science curriculum is being manipulated by central government, for any syllabus that does not fulfil this criterion cannot be certificated by an examining board. In this way, all children taking a science subject at GCSE are forced to face such issues of the day.

2. Differentiated assessment techniques will be employed. The aim of the criterion is to ensure that all candidates are given the opportunity to demonstrate what they *can* do, rather than what they *cannot* do. The emphasis is on success rather than on failure, or, in the words of the criterion, the examination will enable children 'to demonstrate their competence'.

3. Criteria-related grades will be introduced. This means that candidates will be awarded certification upon the basis of achieved personal standards, and not upon comparison with what other children can or cannot do.

4. Whilst the old CSE and O level examinations together were designed to take account of the top 60 per cent of the ability range of the nation's children, the new GCSE examination is 'designed, not for any particular proportion of the ability range, but for *all candidates*, whatever their ability relative to other candidates, *who are able to reach the standards required for the award of particular grades*' (DES, 1985d; our emphasis). This is an important statement, for it belies the commonly held belief that the GCSE is an examination designed for 'all children'. It is not. Indeed, Sir Keith Joseph was quite specific that the objective of the new examination was to bring 'the level of attainment of at least 80 to 90 per cent of all pupils up to at least the level currently associated with the average as reflected in C.S.E. grade 4'. Up to 20 per cent of children (a figure exactly reflected in the Warnock 'one-in-five') are probably not suited for examination at 16+, at least under the GCSE umbrella.

Even for those children who *will* be taking GCSE examinations in science, the national criteria for science subjects do not preclude innovation of content and method, provided the criteria are met. Increased attention to practical work (in the way we have defined it in chapter 6), and the relationship between science and the world of work can be examined through a variety of procedures. Mode 1 examinations (where an examining board both sets and examines the syllabus) will probably continue to be the most popular overall format. In order to take full advantage of the issues discussed in this book for teaching science to less able and underachieving children, serious consideration should be given by working parties of teachers to the potential inherent in Mode 2 (syllabus designed by the school and examined by a board) and Mode 3 (syllabus both designed and examined by the school) alternatives.

The point that we are at pains to make is that the national criteria are far from being totally prescriptive of syllabus content and method, and that the GCSE structure actively encourages curricular initiatives. We believe that the move towards experimental learning espoused in this book and that characterises many Technical and Vocational Education Initiative schemes is possible within present GCSE structures.

DEVELOPMENTS IN SCIENCE EDUCATION RELEVANT TO LOW AND UNDERACHIEVING CHILDREN

Throughout this book we have stressed that a major contributing factor to the successful science education of less able and underachieving children lies in the impetus provided by the class teacher. It must not be the case that these children are left to be taught by the least successful and least motivated teachers in the staffroom. A mark of energetic and committed teachers is their involvement in curricular development, in attempting to tackle some of the special problems that these children have.

One of the philosophies underpinning the organisation of the Secondary Science Curriculum Review (SSCR) is reflected in its insistence upon the 'periphery to centre' model of curriculum development and dissemination (SSCR, 1983). Teachers from local schools have formed some 270 working parties to identify specific science curriculum problems peculiar to their area and have been working together to produce and test materials. The SSCR is currently engaged in reviewing the materials that have been produced over the last three years in science and special education; a publication is due from them early in 1987.

A number of initiatives are also taking place outside the auspices of the SSCR. Of particular note is a science scheme for the less able devised at Gwent College of Higher Education and, at the time of writing, still under extensive trial. It forms an excellent basis for the non-specialist science teacher: the worksheets are easy to read, and can be completed in a variety of ways reminiscent of DARTs activities. There is a wide choice of experiments available related to everyday life and objects – for instance, the use of cabbage water as an indicator for common household acids and bases.

David Page is currently directing a Learning Difficulties in Science Project, which is funded by Cambridge LEA and is based at the Peterborough Educational Centre. The project has a useful 'Action Register' of authorities involved in initiatives in the field.

In the preparation of this book, every LEA in England, Northern Ireland and Wales was contacted to obtain information on current activities relevant to science for low and underachieving children: 69 returns were received, all indicating either current activity (23) or active interest (46). Although many of the science advisers were not themselves spearheading initiatives, a lot of good work was reported at school level. (See Appendix for a list of LEAs that have indicated their willingness to communicate their work with interested people.)

A CHECKLIST FOR CURRICULUM DEVELOPERS

Finally, we offer a distillation of the theoretical issues outlined in preceding chapters. They are presented in some sort of hierarchy, with more general issues appearing first, and more practical considerations towards the end. This is not meant to imply that certain of the checks are more important than others. The stress placed upon any particular check will depend upon the special needs of the children for whom the course work is being prepared.

1. Consider the *aims* of the course. These will vary according to the nature of the problems faced; for example, physical handicap; learning difficulties; demotivation; deprivation of social background; lack of home support, and so on; and most likely some combination of one or more of these. Attention must always be paid to the encouragement of children's improved attitudes to themselves, to others, and to the world of nature.
2. A long-term view will envisage the *development* of attitudes, skills and knowledge in these children. Individualisation is the key. Consider the innate ability (stage of development) of the children, and then their personal interests and inclinations; for example, the differences between boys and girls. We have argued that low achievers in science are nearly always underachievers, and that this is due to a large number of both intrinsic and extrinsic factors. We can expect these children to exhibit an even greater heterogeneity in terms of their 'alternative frameworks' about science than their more able peers, which makes the role of individualisation even more crucial.
3. Challenge the *assumptions* underlying so much of secondary education. Must your course accede to the stricture of 40- or 80-minute lessons? Can the children be taught somewhere other than in science laboratories, so often associated with 'difficult' school work (i.e. 'failure')? Can the children be withdrawn from 'mixed ability' classes? Can a specialist room be developed for these children? Is it necessary that the science specialists teach these children? Might it not be better to adopt the primary school model for at least a proportion of the week, i.e. one teacher teaching across subject boundary lines? What

is meant by a genuinely negotiated science syllabus? What residential trips are available? Can work be done outside normal school hours; for instance, a module of 'night science' (photographing lunar and planetry motion, setting moth traps at dusk, night mammal, bird and insect behaviour)?

4. Never underestimate the *potential* of less able and under-achieving children. They can recognise problems and make sensible searches for relevant solutions. They can often take responsibility for their own learning when that responsibility is given to them. This principle should be linked with the next check.

5. Consider the employment of new and different methods of teaching; for example, using less able and underachieving children as teachers of younger more able children; all kinds of project work; the use of CAL and word processing in particular.

6. Employ rigorous but relevant assessment and recording techniques. Relevant assessment and recording devices make for informed individualised teaching programmes. Nothing succeeds like success. Correctly pitched work permits and encourages success, which needs to be reinforced with records accessible to the children. What about self-assessment? Assessment is necessary in order to remind the busy teacher of individual strengths and weaknesses in the children, so that suitable intervention strategies can be employed as and when they are required. Remember that 'failure' on the part of a child can be due to one or more of three major factors – factors within the child, factors within the teacher, factors within the course material. Figure 10.1 suggests one possible recording device for locating these causes, and is meant for use in mixed ability classes.

7. Select content that, for any reason, has inherent interest for the children, and do not be conditioned by traditional school science. The bending of light by mirrors and lenses is not as significant as the accident at Chernobyl power station, as important as defrosting the Christmas turkey, or as fascinating to most children as using Halley's comet as a form of spaceship for carrying man or devices into deep space and returning them 76 years later. Modular courses permit negotiation and flexibility.

8. Keep up to date with developments in science education (see above). We have exemplified the use of DARTs materials, different styles of writing, the use of 'wait time', and so on. Be aware of, and make use of available facilities and help, e.g. science advisers, SEMERCs and so on.

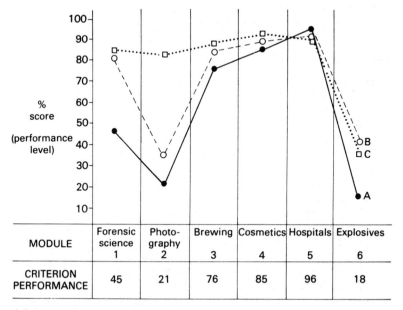

MODULE	Forensic science 1	Photo-graphy 2	Brewing 3	Cosmetics 4	Hospitals 5	Explosives 6
CRITERION PERFORMANCE	45	21	76	85	96	18

A Progress of individual child ●————————————●
B Average progress of class containing child A ○— — — — — —○
C Average progress of all classes in the year group □· ················□

The hypothetical child, A, has scored well below the 80% or so required for mastery learning on three modules – Forensic science, Photography and Explosives.

On module 1, Forensic science, the child has a performance level of 45. Class and year averages are around the 80 mark. It is likely that reasons for poor achievement lie within the child – for example, illness or problems at home?

In module 2, Photography, not only is A scoring badly, but so is the whole class. Yet the average for the whole year is near mastery level. In a mixed ability situation it would seem likely that the teacher is at fault – for example, illness or problems at home?

In module 6, Explosives, all the children throughout the year group perform poorly. Not all children and all teachers are likely to be at fault. The fault is probably with the topic. Maybe concepts of energy change are too abstract for these children at this stage of mental development.

Figure 10.1 A recording device for isolating the main reason(s) for child 'failure' in mixed ability classes

In conclusion, we would reiterate that an entirely inappropriate response to the challenge of teaching science to less able and underachieving children would be to consider content *per se* as more important than, or even as important as, any other of the issues outlined above.

Appendix of resources

Resources available to teachers of less able and poorly motivated children are not easy to find. Traditionally, in science as in most other subjects in the school curriculum, there has been a tendency to modify CSE and O level materials by the more or less judicious selection of the 'easier' sections. It is only very recently that specialist materials for less able children have been produced with the same rigour that has characterised the production of the best examination materials. A good example of this is the TAPS (Bryce *et al.*, 1983) materials to which reference has already been made (pp. 157–8). The consensus amongst those teachers with whom we have worked is that materials written for average and above-average younger children (7–11) are not usually successful with teenaged children. 'Less able' though they may be, they realise that making cotton reel tractors or invisible ink is more suited to little children than to young adults!

The great majority of the resources listed have been produced explicitly for low and underachieving children, and comment is only made when this is not the case. Some of the books and courses available for the average child are mentioned by Travers (1984). The resources appear under four headings:

- professionally published books, worksheets and workcards
- computer assisted learning
- useful addresses
- current initiatives by LEAs.

PROFESSIONALLY PUBLISHED WRITTEN MATERIALS

Low and underachieving children rarely enjoy reading. Although a great deal of the material listed takes this into account, much of it is of more use to the teacher as a source of ideas than it is directly to the children. It is, in our opinion, unlikely that many of the materials listed in this section are suitable for entire class distribution. More valuable would be their inclusion into a laboratory library, where they are accessible to the children when required. The materials are listed here in alphabetical order of title. Full details of books are given in the References under author name.

Resources intended for teachers

Title & Author	Notes
A Safety Handbook for Science Teachers, by Everett & Jenkins (1980).	Especially valuable to teachers who are not science specialists; now in its third edition.
School Science Laboratories: A Handbook of Design, Management & Organization, by Archenhold, Jenkins & Wood-Robinson (1978).	Packed with useful ideas and information.
Teaching Pupils of Low Scientific Attainment, by Foster (ed.) (1984).	Only 22pp, but some good hints about organising a specialist laboratory for the less able. Still just about available.
Urban Ecology, by Collins (1986).	A resource of ideas for teaching materials available in the town or the school grounds. Section three contains a collection of projects possible in the urban environment.

Materials written for children

Title & Author	Notes
Access to Science, by Mitchell & Snape (1982)	Practical science for children with a reading age of about 9 years. Six pupils' books on Measurement, Classification, Separating Substances, Water, Reproduction, and Air. Teacher guides contain exercises to be photocopied for class use.

Title & Author	Notes
Basic Skills in Science, by Merrigan & Herbert (1979)	Reviewed in *School Science Review* as 'an outstanding contribution'. 35 skills are identified as essential for the lowest 30 per cent of the ability range in lower secondary.
Child Care and Development, by Minett (1986)	Comprehensive and well illustrated. Topics are self-contained for flexibility of use; focus on the role of the family.
Databank, by Crystal & Foster (1985)	Some 30 books in the series, each only 24pp long, and usable in project work. Free worksheets. Topics include Heat, Light, Sound, Volcanoes, and Fishing.
Fit for Life, by McNaughton (1986)	Health education unit for 5–16 year olds; levels 2 and 3 are suitable for secondary children. Latter includes 116 photocopy masters. Free sample pack.
Insight to Science, by Inner London Education Authority Curriculum Development team (1978)	Written for children of all abilities, 11–13 years. Somewhat dated now, the individual work cards are plain but usefully graded for difficulty.
Introducing Science, by Jackson (1985)	Series originally written for primary children, with the aim of improving attitudes to science. 12 books, 6 teacher guides on the usual topics, e.g. Energy, Light, Heat, Electricity, etc.
It's Your Life, by Smith & Curtis (1986)	Eight problem areas for youngsters to think about and discuss: babies and parents,

Title & Author	*Notes*
	sex roles, sex and birth control, race prejudice, drinking, smoking (but not drugs), etc. Cartoons are highly professional.
LAMP project, by Bowers (ed.) (1976) and Wilkinson & Bowers (1976)	One of the earlier attempts to design a science curriculum for the less able based on educational theory. It provides a large and useful set of 16 topics and 2 teacher handbooks, including Pollution, Photography, Gardening, and Flight.
Learning through Science, Schools Council (1980)	Can be used in conjunction with *Science 5–13* (see below), but essentially for primary children.
Modular Science, by Fairbrother, Jenkins & Scott (eds) (1985)	A series of some 20 booklets, each of 32pp with associated teacher notes. It provides materials from which teachers can construct their own course. Includes topics like Conserve or...?, The Commonest Compound, Space Science, Plastics Everywhere, etc.
Physics Around Us, by Gardner & Scott (1984)	Just 48pp, the book covers the key topic areas of most physics courses. For the 14–16 age range.
Reading About Science, by Kellington (ed.) 1982)	Highly recommended series of five books. Each chapter covers just 2pp, and is graded for readability and content. Some of the questions could be more DARTs-oriented – a job for the teacher!

Title & Author	*Notes*
Science at Work, by Taylor (ed.) (1979)	This series has been on the market for some time, but is still extensively used. Some fascinating titles, e.g. You & Your Mind, Forensic Science, Cosmetics. 18 units and teacher guides.
Science 5–13, Schools Council (1973)	Not written specifically for the less able, but the series is highly organised in terms of objectives and investigations, which are related to stages in children's cognitive development. 27 units include Coloured Things, Holes, Gaps & Cavities, and Time.
Science for Children with Learning Difficulties, Schools Council (1983)	Very much primary oriented, and cutting out paper clothes is unsuitable material for most young adults. But some of the material is adaptable, e.g. recording shadows.
Science for the Individual, Resources for Learning Development Unit (1985)	The project aims to provide a large collection of resources suitable for pupils of low academic achievement. The bank is not designed as a course. Units include Building Science, Body Maintenance, and Grub Up! Order forms from RLDU.
Science Horizons, by Hudson & Slack (1985)	Designed for use by non-specialist teachers, but the materials are often primary oriented. It is possible to purchase units in just the 10–12 age range (level 2b), and these include Understanding the Weather, Shipshape, and Pond Life. A free evaluation

Title & Author	*Notes*
	pack is available from the publishers.
Scientific Eye, by Hart-Davis (1986)	Not specifically for the less able, but well presented. Each of the 59 chapters covers a double-page spread and explores one fundamental concept. It is also a TV series. Useful source material that will require some adaptation.
Start with Science, by Dixon (1986)	Again, not specifically written for the less able. Packs contain 50 photocopy masters for worksheets on such topics as Light, Air, Water, and Plants. Free sample material available from publisher.
Steps in Science, by Bateman & Lidstone (1986)	Nicely presented with good photography and professional cartoons. Designed to augment practical sessions. Low readability, the three books cover work for 11–14 year olds.
Techniques for the Assessment of Practical Skills in Foundation Science (TAPS), by Bryce *et al.* (1983)	A highly sophisticated criterion-referenced programme for the systematic assessment of science practical skills, designed specifically for less able 14–16 year olds. Over 300 test items relate to 46 practical skill objectives, from which comprehensive profiles of achievement can be compiled. The pack is expensive (£125), but complete. An invaluable departmental resource, packed with good ideas, all of which have been extensively tested

Title & Author	*Notes*
	in schools. Designed for use with the Scottish Foundation Course, but highly adaptable.
This is Science, by Dobson (1985)	Integrated science for average and below-average 11–13 year olds in mixed ability classes.

COMPUTER ASSISTED LEARNING

Books, magazines and newsletters for teachers

Behrmann, M. (1985), *Handbook of Microcomputers in Special Education*
Green, F. *et al.* (1982), *Microcomputers in Special Education*
Hawkridge, P., Vincent, T., & Hales, G. (1984), *New Information Technology in the Education of Disabled Children and Adults*
Hawthorne, P. (1985), *The Science Teacher's Companion to the BBC Microcomputer*
Hogg, R. (1984), *Microcomputers and Special Educational Needs*
Hope, M. (ed.) (1986), *The Magic of the Micro: a Resource for Children with Learning Difficulties*
Learning to Cope: a magazine published by Educational Computing (11 issues per annum) (see Address List)
Microelectronics and Children with Special Educational Needs: a free newsletter – contact R. A. Fitt (see Address List)
Rostron, A., & Sewell, D. (n.d.), *Microtechnology in Special Education*
SEMERC regional newsletters: contact your local SEMERC (see Address List)
Sparkes, R. A. (1983), *The BBC Microcomputer in Science Teaching*
Taber, F. M. (1983), *Microcomputers in Special Education*
Wilkinson, A. L. (1983), *Classroom Computers and Cognitive Science*

Hardware

Aids to Communication in Education (ACE) welcomes direct approaches from parents, children and all professionals associated with children who have physical impairments and communication difficulties. The centre has many devices to match users to available machines, including special interfaces, single and multiple switches, microwriters, special keyboards and light sensitive devices, voice synthesisers and remote-controlled devices such as Turtle. ACE is hoping to establish a detailed database providing a hardware and software inventory, bibliography, and a mailing list of contacts.

Concept keyboard, which has been described in chapter 6. A most versatile piece of hardware, it is on display in SEMERCs throughout the country, and is manufactured by A & B European Marketing. The A4 pack costs £99.50 including VAT, the A3 version, £127. The packs include keyboard, interface lead, introductory overlay set, user guide and four items of software – DC Meters, Heat the House, Small Mammal Survey, and Gumshoe Logic.

Expanded keyboard for BBC microcomputer, from Special Technology Ltd. The keyboard is an enlarged and simplified QWERTY version. It has built-in delays, which largely eliminate the effect of unintended key presses, and gives up to two seconds to release the key before repeating starts. Other advantages for the less able are the recessed keys, which reduce accidental presses, and the SHIFT and BREAK and CONTROL and BREAK operations can be performed with a single finger. £243 including interface, postage, insurance and packing, but excluding VAT.

Software

There is a vast array of software on the market. It falls into two main categories, that written for average and most able children, and that written especially for children with learning difficulties. Unlike the equivalent written materials, much software written for average children is suitable for the less able, since the complexity of the response made by the computer depends upon the complexity of the input from the child. One particularly encouraging sign is that one publishing house, Cambridge Microsoftware, now offers an evaluation service, analogous to the inspection service offered to purchasers of textbooks.

Look for full use of the special advantages offered by software over textbooks and written materials – in particular, personalisation, interaction between child and computer, and good use of colour and graphics. Some programs allow children to make their own pictorial representations of data (e.g. Picfile, from Cambridge Microsoftware). Philip Harris Biological produce some nice interactive simulations, e.g. on the use of controls, and some playable science games are available, e.g. the blood circulation game. Some exciting new materials are being developed by Ken Turner at the Department of Education in the University of Cambridge. They will be particularly useful in health education, and enable children to be connected to a BBC microcomputer where the effect of exercise on heart beat and other body functions can be seen directly on a VDU, and recorded on a printer. Thus children will have their own individual printouts for inclusion in their files. Publication is due in 1987.

Software catalogues should be obtained from:

Arnold Publishers Ltd (including programs produced by the Chelsea Science Simulation Project)
Audio Visual Productions (this company has a large stock of videos and other visual aids)
Cambridge Microsoftware (including their primary software)
Griffin and George
Heinemann Computers in Education
Hutchinson Software
Longman Microsoftware (including programs produced by the Schools Council and Warwick Science Simluations)
Microspecial (Scottish Education Department)
Philip Harris Biological

The Interactive Videos in Schools Project is also geared to the use of interactive video by children with special needs.

The regional Special Education Microelectronic Resources Centres (SEMERCs) are in business solely for the needs of children requiring special education. They can be approached directly, or better still through the many co-ordinators in the four regional areas. Write to your regional SEMERC for a list of names and addresses of your local co-ordinator. SEMERC supplies inventories of available software, most of which is free, and is constantly producing new materials. In particular, ask for the briefing leaflets, which are also available from the Council for Educational Technology (CET). Briefing No. 2, for example, provides a core library of programs available for the BBC micro, and includes 'Flowers of Crystal' and 'Eating for Health'. Briefing No. 14 gives a comprehensive list of suppliers of software and peripherals suited to children with special educational needs. Framework programs, such as 'Touch Explorer', for use with the concept keyboard are also available from SEMERC.

ADDRESS LIST

The majority of the addresses given below relate to personnel, companies and institutions referred to elsewhere in this book. That is, they are specifically relevant to the teaching of science to low and underachieving children and, in a few cases, to the physically handicapped. A more comprehensive list of addresses for science teachers generally is available from the Centre for Studies in Science and Mathematics Education at the University of Leeds, in a booklet called 'Addresses for Science Teachers'.

A & B European Marketing, Forrest Farm Industrial Estate, Whitchurch, Cardiff CF 7YS (0222 618336)

Aids to Communication in Education (ACE) Centre, Ormerod School, Waynflete Rd, Headington, Oxford OX3 8DD (0865 63508) 9am to 5pm during term time

Arnold Publishers Ltd, Woodlands Park Avenue, Maidenhead, Berks

Association for Science Education (ASE), College Lane, Hatfield, Herts AL10 9AA

Audio Visual Productions, Hocker Hill House, Chepstow, NP6 5ER

Cambridge Microsoftware, Home Sales Department, Cambridge University Press, The Edinburgh Building, Shaftesbury Rd, Cambridge CB2 2RU

Centre for Studies in Science and Mathematics Education, University of Leeds, Leeds

Council for Educational Technology (CET), 3 Devonshire Street, London W1N 2BA

Educational Computing, Priory Court, 30–32 Farringdon Lane, London EC1R 3AU

Griffin and George, Bishop Meadow Road, Loughborough, Leicestershire LE11 0RG

Gwent College of Higher Education, Caerleon, Newport, Gwent

Heinemann Computers in Education Ltd, Freepost, EM 17, 22 Bedford Square, London WC1B 3HH

Hutchinson Software, 17–21 Conway Street, London W1P 5HL

Interactive Videos in Schools Project, 27 Marylebone Road, London NW1 5JS (01 935 5488)

Longman Microsoftware, 33–35 Tanner Row, York YO1 1JP

Microelectronics and Children with Special Educational Needs. Free newsletter from Mr R. A. Fitt, 2 Monkseaton Road, Sutton Coldfield, West Midlands B72 1LB

Microspecial, Collins Soft/Hill MacGibbon, 8 Grafton Street, London W1X 3LA (01 493 7070)

Philip Harris Biological Ltd, Oldmixon, Weston-Super-Mare, Avon BS24 9BJ

Research Centre for the Education of the Visually Handicapped, Selly Wick House, 59 Selly Wick Road, Birmingham B29 7JE

Resources for Learning Development Unit (RLDU), Bishop Road, Bishopston, Bristol BS7 8LS (0272 428208)

SEMERCs:

Bristol SEMERC, Bristol Polytechnic, Redland Hill, Bristol BS6 6U2 (0272 733141)

Manchester SEMERC, Manchester Polytechnic, Hathersage Road, Manchester M13 0JA (061 225 9054, ext. 284)

Newcastle SEMERC, Newcastle Polytechnic, Coach Lane campus, Newcastle-upon-Tyne, NE7 7XA (0912 665057)

Redbridge SEMERC, The Dane Centre, Melbourne Road, Ilford, Essex IG1 4HT (01 478 6363)

Secondary Science Curriculum Review (SSCR), School of Education, University of Bath, Claverton Down, Bath BA2 7AY (Regional Project Leader: Christine Ditchfield)

Special Technology Ltd, Freepost, Southport, Merseyside PR8 1BR

K. O. Turner, Resources Centre, 17 Brookside, Cambridge CB2 1JG

WATCH Trust for Environmental Education, High Beach, Loughton, Essex IG10 4AF.

CURRENT INITIATIVES BY LEAs

In each case, it is the science adviser (whose name appears in parentheses) who should be contacted in the first instance.

Avon (Dr P. Armitage, senior adviser science and technology)   A working party of science teachers from a number of schools has published its own booklet, *Science for Pupils with Special Educational Needs* (1985). This identifies the skills, concepts, and attitudes to science that all children are deemed to need, gives suggestions for activities, reviews current published materials, lists local specialist facilities such as the Avon Wildlife Trust, the RLDU, and the Bristol Museum, and gives details of the Science Certificate issued by the authority to children who are able to demonstrate a list of skills in science.

Berkshire (Dr R. C. Dorrance)   A group of teachers has been working under the auspices of the SSCR. Some of the work is currently undergoing trials in Berkshire schools, e.g. 'Heat', and 'The Environment'.

Birmingham (P. C. Rudge)   Whilst the authority has not produced materials centrally, the adviser can put enquirers in touch with schools with active schemes.

Cheshire (B. L. Leek)   The authority has produced a total of 12 units, some currently undergoing trials. For instance, 'UFO Crash' is a 22-page booklet that begins with a newspaper article purported to be taken from the *Guardian* and entitled 'UFO Invasion'. The children are provided with test kits, and are expected to visit the site of the crash to collect samples for testing. For instance, the density kit enables them to discover the density of a number of metals found on the site, which can then be identified from a density chart. But what to do with the metal of unknown composition that does not appear on the chart of all known earth metals...?

Cleveland (J. A. Slade)
Durham (J. H. Crossland)

Enfield, London Borough of (Jackie Hardie)   The bulk of the work in this authority has been developed on 'Environmental Science'.

Gateshead (I. Hills)

Hertfordshire (C. P. E. Short)   Two projects have been particularly successful. The Watford Industry Project is run by a group of schools in Watford for non-CSE pupils, and the Herts Achievement Project has devised a profile-reported certificate for pupils in the lowest ability group.

Lancashire (P. H. Gardner)

Mid Glamorgan (C. K. Rawlins and P. Pearson)   Most of the work has been directed at primary schools, e.g. topic cards such as 'Measuring Shadows' and 'Snails on the Move'. Work has also been completed on photography and the use of living materials in classrooms.

Nottinghamshire (J. McLaren)   A technology initiative for primary and lower secondary schools.

Redbridge, London Borough of (Dr Valerie Tracey)   Recently Dr Tracey directed a DES Regional Course on Science for Secondary School Pupils with Special Needs. The course handbook contains examples of units compiled by teachers in both special and ordinary schools in the London Boroughs of Enfield, Haringey, Redbridge, and Waltham Forest, and in Essex and Herts.

Suffolk (L. J. Smith)

West Glamorgan (I. Page)   At least one school has produced its own scheme. This will be rewritten and trialled by a group of schools for dissemination within the authority.

Wolverhampton (C. Edwards)

References

Abercrombie, K. (1968) Paralanguage, *British Journal of Communication*, **3**, 55–59

Aikenhead, G. S. (1985) 'Collective social decision-making: implications for teaching science', in D. Gosling and B. Musschenga, *Science Education and Ethical Values*. Geneva: WCC Publications, 55–67

Allen, V. L. (ed.) (1976) *Children as Teachers: Theory and Research on Tutoring*. New York: Academic Press

Amidon, E. J., and Hough, J. B. (eds) (1967) *Interaction Analysis: Theory, Research and Application*. Reading, MA: Addison-Wesley

APU [Assessment of Performance Unit] (1985) *Science at Age 15*. London: HMSO

Archenhold, W. F., Jenkins, E. W., and Wood-Robinson, C. (1978) *School Science Laboratories: A Handbook of Design, Management and Organization*. London: Murray

Argyle, M. (1975) *Bodily Communication*. London: Methuen

ASE [Association for Science Education] (1979a) *Alternatives for Science Education*. Hatfield, Herts: ASE

ASE (1979b) *Resource-based Learning in Science*. Hatfield, Herts: ASE

ASE (1980) *Language in Science: Study series no. 16*. Hatfield, Herts: ASE

ASE (1981) *Education Through Science*. Hatfield, Herts: ASE

ASE (1984) *Rethinking Science?* Hatfield, Herts: ASE

ASE (1985) *Planning for Science in the Curriculum*. Hatfield, Herts: ASE

Ausubel, D. P. (1968) *Educational Psychology: A Cognitive View*. New York: Holt, Rinehart

Ausubel, D. P. (1978) *The Psychology of Meaningful Verbal Learning*. New York: Grune & Stratton

Avon (County of) (1985) *Science for Pupils with Special Educational Needs*. Avon LEA

Ayres, D. G. and Hinton, R. A. L. (1985) Meeting the needs of visually handicapped students in ecology, *School Science Review*, **67**, 18–26

Bantock, G. H. (1971) Towards a theory of popular education, *The Times Educational Supplement*, 12 and 19 March

Barlex, D. and Carré, C. (1985) *Visual Communication in Science: Learning through Sharing Images*. Cambridge: Cambridge University Press

Barnes, D. (1969) *Language, the Learner and the School*. Harmondsworth, Middx: Penguin

Barnes, D. (1976) *From Communication to Curriculum*. Harmondsworth, Middx: Penguin

Barnes, D. (1982) *Practical Curriculum Study*. London: Routledge & Kegan Paul

Bateman, R. and Lidstone, P. (1986) *Steps in Science*. London: Hutchinson

Beatty, J. W. and Woolnough, B. E. (1982a) Practical work in 11–13 science: the context, type and aims of current practice, *British Journal of Research in Education*, **8**, 23–30

Beatty, J. W. and Woolnough, B. E. (1982b) Why do practical work in 11–13 science? *School Science Review*, **64**, 768–770

Behrmann, M. (1985) *Handbook of Microcomputers in Special Education*. Windsor: NFER/Nelson

Benn, C. and Simon, B. (1970) *Half Way There: Report on the British Comprehensive School Reform*. Harmondsworth, Middx: Penguin

Benton, P. (1981) 'Writing: how it is received', in C. Sutton (ed.) *Communication in the Classroom*. London: Hodder & Stoughton

Bernstein, B. (1973) *Class, Codes and Control: Vol 1 – Theoretical Studies Towards a Sociology of Language*. London: Routledge & Kegan Paul

Beswick, N. (1972) *School Resource Centres*. London: Evans/Methuen

Beveridge, M. (1985) The development of young children's understanding of the process of evaporation, *Br. J. Educ. Psychol.*, **55**, 84–90

Black, H. D. and Dockrell, W. B. (1984) *Criterion Referenced Assessment in the Classroom*. Edinburgh: The Scottish Council for Research in Education

Black, P. and Ogborn, J. (1981) 'Science in the school curriculum', in J. White *et al.*, *No Minister: A Critique of the DES Paper 'The School Curriculum'*. Bedford Way Papers No. 4, London: Heinemann Educational Books

Blaisdell, A. F. (1897) *A Practical Physiology: A Textbook for Higher Schools*. Boston, MA: Ginn & Co

Bloom, B. S. (ed.), (1956) *Taxonomy of Educational Objectives. Handbook I: Cognitive Domain*. New York: David McKay & Co.

Board of Education (1943) *Report of the Committee of the Secondary Schools Examination Council on Curriculum and Examinations in Secondary Schools* (The Norwood Report). London: HMSO

Bock, M. (1983) 'The influence of pictures on processing texts: reading time, intelligibility, recall, aesthetic effect, need for re-reading', in G. Rickheit and M. Bock (eds) *Psycholinguistic Studies in Language Processing*. Berlin: de Gruyter

Bowers, J. (ed.) (1976) *Less Academically Motivated Pupils (LAMP) Project*. Hatfield, Herts: Association for Science Education

Brandt, G., Turner, S., and Turner, T. (1985) 'Science education in a multicultural society'. Report on a conference held at the University of London Institute of Education

Brennan, W. (1979) *Curricular Needs for Slow Learners*. London: Evans/Methuen

Broadbridge, J. (1986) 'Children's perceptions of depth in pictures'. DASE dissertation, University of Manchester

Broadfoot, P. (1979) *Assessment, Schools and Society*. London: Methuen

Brown, S. (1981) *What Do They Know? A Review of Criterion Referenced Assessment*. Edinburgh: HMSO

Bruner, J. S. (1960) *The Process of Education*. Chicago, IL: University of Chicago Press

Bruner, J. S. (1972) *The Relevance of Education*. Harmondsworth, Middx: Penguin

Brush, L. (1979) Avoidance of science and stereotypes of scientists, *Journal of Research in Science Teaching*. **16**, 237–241

Bryce, T., McCall, J., MacGregor, J., Robertson, I. J., and Weston, R. A. J. (1983) *Techniques for the Assessment of Practical Skills in Foundation Science, Teachers' Guide*. London: Heinemann

Burns, J. C., Okey, J. R., and Wise, K. C. (1985) Development of an integrated process skill test: TIPS II, *Journal of Research in Science Teaching*, **22**, 169–177

Carré, C. (1981) *Language Teaching and Learning: Science*. London: Ward Lock Educational

Carrick, T. (1977) Comparison of recently published Biology textbooks for first examinations, *Journal of Biological Education*. **11**, 163–175

Carrick, T. (1982) More new textbooks for first examinations in Biology, *Journal of Biological Education*. **16**, 253–264

Cassells, J. R. T. (1980) Language and learning: a chemist's view, *Teaching English*, **14**, 24–27

Cassells, J. R. T. and Johnstone, A. H. (1983) The meaning of words and the chemistry of teaching, *Educ. Chem.*, **20**, 10–11

Christie, T. (1986) 'Assessing children's literacy: a Joint Matriculation Board funded project'. Paper given to the Findlay Society, January 1986, University of Manchester

Clegg, A. S. and Morley, M. (1980) Applied science – a course for pupils of low educational achievement, *School Science Review*, **61**, 454

Clift, P. (1982) LEA schemes for self evaluation: a critique, *Education Research*, **24**, 262–271

CLIS [Children's Learning in Science Project] (1984a) 'Aspects of secondary students' understanding of heat'. Leeds: Centre for Studies in Science and Mathematics Education, Leeds University

CLIS (1984b) 'Aspects of secondary students' understanding of plant nutrition: full report'. Leeds: Centre for Studies in Science and Mathematics Education, Leeds University

CLIS (1984c) 'Aspects of secondary students' understanding of energy'. Leeds: Centre for Studies in Science and Mathematics Education, Leeds University

Cockroft, W. H. (1982) *Mathematics Counts. Report of the Committee of Enquiry into the Teaching of Mathematics in Schools*. London: HMSO

Collins, M. (1986) *Urban Ecology*. Cambridge: Cambridge University Press

Cooper, K. (1976) 'Curriculum evaluation – definition and boundaries', in D. Tawney (ed.) *Curriculum Evaluation Today: Trends and Implications*. London: Macmillan/Schools Council, 1–10. Edinburgh: Oliver & Boyd

Crystal, D. and Foster, J. L. (1985) *Databank*. London: Arnold

Dark, H. G. N., Egren, J. R., Garner, P. H., and McGregor, D. W. (1985) A model for curriculum development in local education authorities, *School Science Review*, **66**, 634–644

Davies, F. and Greene, T. (1984) *Reading for Learning in the Sciences*. Edinburgh: Oliver & Boyd

Deale, R. N. (1975) *Assessment and Testing in the Secondary School*, Schools Council Exam. Bull. No. 32. London: Evans/Methuen

Delaney, M. R. (1966) *Build Your Body: A Practical Series of Cut-out and Build-up Models of Organs of the Body*. London: Macmillan

DES [Department of Education and Science] (1963) *Half Our Future: A Report of the Central Advisory Council for Education (England)* (The Newsom Report). London: HMSO

DES (1977) *Curriculum 11–16*. London: HMSO

DES (1978a) *Special Educational Needs: Report of the Committee of Enquiry into the Education of Handicapped Children and Young People* (The Warnock Report). London: HMSO

DES (1978b) *Mixed Ability Work in Comprehensive Schools*. London: HMSO

DES (1981a) *The School Curriculum*. London: HMSO

DES (1981b) *Education Act 1981*, Circular No 8/81. London: HMSO

DES (1982) *Science Education in Schools*. London: HMSO

DES (1984) *Education Observed 2*. London: HMSO

DES (1985a) *Science 5–16*. London: HMSO

DES (1985b) *Better Schools*. London: HMSO

DES (1985c) *Education Observed 3: Good Teachers*. London: HMSO

DES (1985d) *General Certificate of Secondary Education: A General Introduction*. London: HMSO

DES (1985e) *The Curriculum from 5 to 16*. London: HMSO

Dewey, J. (1916) *Democracy and Education*. New York: The Free Press

Dibbs, D. R. (1982) 'An investigation into the nature and consequences of teachers' implicit philosophies of science'. PhD thesis, University of Aston

Dillashaw, F. G. and Okey, J. R. (1980) Test of integrated process skills for secondary science students, *Science Education*, **64**, 601–608

Dixon, M. (1986) *Start with Science*. London: Arnold

Dobson, K. (1985) *This Is Science*. Basingstoke, Hants: Macmillan

Douglas, J. W. B. (1964) *The Home and the School*. London: MacGibbon & Kee

Driver, R. (1983) *The Pupil as Scientist?* Milton Keynes: Open University Press

Dweck, C. (1984) The power of negative thinking, *The Times Educational Supplement*, 21 September

Dyke, R. G. (1986) 'Touch explorer: a simple framework program supporting classroom activity'. Manchester: Special Education Microelectronic Resources Centre

Ebel, R. L. (1972) *Essentials of Educational Measurement*. Englewood Cliffs, NJ: Prentice-Hall

Edwards, A. D. (1976) *Language in Culture and Class*. London: Heinemann Educational Books

Edwards, A. D. and Furlong, V. J. (1978) *The Language of Teaching*. London: Heinemann

Eggleston, J. F., Galton, M. J., and Jones, M. E. (1976) *Processes and Products of Science Teaching*. London: Macmillan/Schools Council

Eisner, E. W. (1969) 'Instructional and expressive objectives: their formulation and use in curriculum', in W. J. Popham (ed.) *Instructional Objectives*, AERA Monograph No. 3, Chicago, IL: Rand McNally

Everett, K. and Jenkins, E. W. (1980) *A Safety Handbook for Science Teachers*. London: Murray

Fairbrother, B., Jenkins, E. W., and Scott, P. (eds) (1985) *Modular Science*. Glasgow: Blackie

Feidler, F. E. (1967) *A Theory of Leadership Effectiveness*. New York: McGraw-Hill

Fensham, P. J. (1985) Science for all: a reflective essay, *Journal of Curriculum Studies*, **17**, 415–435

Ferry, G. (1985) Was WISE worthwhile? *New Scientist*, No. 1437, 28–31

Fish, J. R. (1982) 'Training for work in ordinary schools. Teacher training and special education'. An invitation conference, Bishop Grosseteste College, Lincoln, 4 and 5 March

Fitz-Gibbon, C. T. (1977) An analysis of the literature on cross-age tutoring, *CSE report on tutoring No. 5*. Los Angeles, CA: UCLA

Fitz-Gibbon, C. T. (1978a) A survey of tutoring projects, *CSE Report on Tutoring No. 118*, Los Angeles, CA: UCLA

Fitz-Gibbon, C. T. (1978b) Setting up and evaluating tutoring projects, *CSE Report No. 122*. Los Angeles, CA: UCLA

Floud, J. E., Halsey, A. H., and Martin, F. M. (1957) *Social Class and Educational Opportunity*. London: Heinemann

Forrest, G. H., Smith, G. A., and Brown, M. H. (1970) *General Studies (Advanced) and Academic Attitudes*. Manchester: Joint Matriculation Board

Foster, D. (1984) *Teaching Pupils of Low Scientific Attainment, Resources for Learning Development Unit*. Bristol: Cecil Powell Centre, University of Bristol

Freeman, J. (1979) *Gifted Children*. Lancaster: MTP Press

Freeman, J. (ed.) (1985) *The Psychology of Gifted Children*. Chichester: Wiley

Frith, D. S. and Macintosh, H. G. (1984) *A Teacher's Guide to Assessment*. Cheltenham: Stanley Thornes

Gardner, D. J., and Scott, B. M. (1984) *Physics Around Us*. London: Arnold

Garrett, J. and Dyke B. (in press) *Microelectronics and Pupils with Special Educational Needs: Getting Started – Some Support Material for the In-Service Training of Teachers*. Manchester: Manchester University Press

Gartner, A., Kohler, M. C. and Riessman, F. (1971) *Children Teach Children: Learning by Teaching*. New York: Harper & Row

Gattegno, C. (1970) *What We Owe Children: The Subordination of Teaching to Learning*. London: Routledge & Kegan Paul

Gauld, C. F. (1982) The scientific attitude and science education: a critical reappraisal, *Science Education*, **66**, 109–121

Goldenberg, E. P., Russell, S. J., and Carter, C. J. (1984) *Computers, Education and Special Needs*. Reading, MA: Addison-Wesley

Goodlad, S. (1979) *Learning by Teaching: An Introduction to Tutoring*. London: Community Service Volunteers

Gould, C. D. (1978) Practical work in sixth form biology, *Journal of Biological Education*, **12**, 33–38

Green, F. et al. (1982) *Microcomputers in Special Education*. London: Schools Council

Gronlund, N. E. (1982) *Constructing Achievement Tests*. Englewood Cliffs, NJ: Prentice-Hall

Guthrie, H. B. (1980) 'A brief review of the literature related to practical work in science'. MEd thesis, University of Bath

Hacker, R. G. (1982) *The Science Lesson Analysis System*. Nedlands, W. Australia: University of Western Australia

Hadden, R. A. and Johnstone, A. H. (1983) Secondary school pupils' attitudes to science: the year of decision, *European Journal of Science Education*, **5**, 429–438

Hall, E. T. (1966) *The Hidden Dimension*. Garden City, NY: Doubleday

Harding, J. (1983) *Switched Off: The Science Education of Girls*. York: Longman/Schools Council

Hargreaves, D. H. (1982) *The Challenge for the Comprehensive School: Culture, Curriculum and Community*. London: Routledge & Kegan Paul

Harlen, W. (1978) Does content matter in primary science? *School Science Review*, **59**, 614–625

Harlen, W. (1983) *Guides to Assessment in Education: Science*. London: Macmillan

Hart-Davis, A. (1986) *Scientific Eye*. London: Bell & Hyman

Harvey, T. J. (1985) Gender differences in attitudes to science and school for first year secondary school children in a variety of teaching groups, *Educational Review*, **37**(3), 281–288

Hawkridge, P., Vincent, T., and Hales, G. (1984) *New Information Technology in the Education of Disabled Children and Adults*. London: Croom Helm

Hawthorne, P. (1985) *The Science Teachers Companion to the BBC Microcomputer*. London: Macmillan

Head, J. J. (1976) *Science Through Biology*. London: Arnold

Head, J. O. (1985) *The Personal Response to Science*. Cambridge: Cambridge University Press

Henerson, M. E., Morris, L. L., and Fitz-Gibbon, C. T. (1978) *How to Measure Attitudes*. Beverley Hills, CA: Sage

Hirst, P. H. (1965) 'Liberal education and the nature of knowledge', in R. D. Archambault (ed.) London: Routledge & Kegan Paul

HMI [HM Inspectorate] (1978) *Aspects of Primary Education in England and Wales*. London: HMSO

HMI (1979) *Aspects of Secondary Education in England*. London: HMSO

HMI (1980) *Girls and Science – Matters for Discussion*, No. 13. London: HMSO

HMI (1984) *Slow Learning and Less Successful Pupils in Secondary Schools: Evidence from Some HMI Visits*. London: HMSO

Hodson, D. (1985a) Philosophy of science, science and science education, *Studies in Science Education*, **12**, 25–27

Hodson, D. (1985b) Profiling: practical problems and teacher anxiety, *Education in Science*, **111**, 41–43

Hodson, D. (1986a) Philosophy of science and science education, *Journal of Philosophy of Education*, **20**, 241–51

Hodson, D. (1986b) The role of assessment in the 'curriculum cycle': a survey of science department practice, *Research in Science & Technological Education*, **4**, 7–17

Hodson, D. (1986c) Rethinking the role and status of observation in school science, *Journal of Curriculum Studies*, **18**, 381–96

Hodson, D. (1986d) A checklist of questions for science curriculum evaluation, *School Science Review*, in press

Hodson, D. (1986e) 'The evaluator's perspective', in R. Sutton (ed.) *Assessment in Secondary Schools: The Manchester Experience.* York: Longman/School Curriculum Development Committee, 63–87 and 103–123

Hodson, D. and Brewster, J. (1985) Towards science profiles, *School Science Review*, **67**, 231–240

Hofstein, A., Mandler, V., Ben-Zvi, R., and Samuel, D. (1980) Teaching objectives in chemistry: a comparison of teachers' and students' priorities, *European Journal of Science Education*, **2**, 61–66

Hogg, R. (1984) *Microcomputers and Special Educational Needs.* Stratford-on-Avon: National Council for Special Education

Hope, M. (ed.) (1986) *The Magic of the Micro – A Resource for Children with Learning Difficulties.* Newcastle upon Tyne: Council for Educational Technology

Hudson, J. and Slack, D. (1985) *Science Horizons.* London: Macmillan

Hull, R. and Adams, H. (1981) *Decisions in the Science Curriculum; Organization and Curriculum.* Hatfield, Herts: Association for Science Education/Schools Council

Hull, R. and Adams, H. (1982) *Studies in Decision-making for Science Education. A Guide for In-service Training.* Hatfield, Herts: Association for Science Education

Hurd, P. D. (1971) *New Directions in Teaching Secondary School Science.* Chicago, IL: Rand McNally

Ichikawa, S. (1981) In situ monitoring with Tradescantia around nuclear power plants, *Environmental Health Perspectives*, **37**, 145–164

Inner London Education Authority (1978) *Insight to Science.* London: Addison-Wesley

Jackson, S. (1985) *Introducing Science.* Glasgow: Blackie

James, C. (1968) *Young Lives at Stake: A Reappraisal of Secondary Schools.* London: Collins

Jenkins, E. W. (ed.) (1973) *The Teaching of Science to Pupils of Low Educational Attainment*. Leeds: Centre for Studies in Science and Mathematics Education, Leeds University

Johnson, J. (1986) 'Improving self-esteem and motivation: an overview', in M. Hope (ed.) *The Magic of the Micro – A Resource for Children with Learning Difficulties*. Newcastle upon Tyne: Council for Education Technology

Johnson Abercrombie, M. L. (1960) *The Anatomy of Judgement*. Harmondsworth, Middx: Pelican Books

Johnstone, A. H., MacGuire, P. R. P., Friel, S., and Morrison, E. W. (1983) Criterion-referenced testing in science – thoughts, worries and suggestions, *School Science Review*, **64**, 626–634

Jones, A. V. (1983) *Science for Handicapped Children*. London: Souvenir Press (Educational and Academic)

Judge, H. (1984) *A Generation of Schooling: English Secondary Schools Since 1944*. Oxford: Oxford University Press

Kellington, S. (ed.) (1982) *Reading About Science*, Books 1–5. London: Heinemann Educational Books

Kellington, S. H. and Mitchell, A. C. (1980) Designing an assessment system on the principles of criterion-referenced measurement, *School Science Review*, **61**, 765–770

Kelly, A. (1979) Where have all the women gone? *Physics Bull.*, **30**, 108

Kelly, A. (1981a) (ed.) *The Missing Half: Girls and Science Education*. Manchester: Manchester University Press

Kelly, A. (1981b) 'Sex differences in science achievement: some results and hypotheses', in A. Kelly (ed.) *The Missing Half*. Manchester: Manchester University Press

Kelly, A. (1985) The construction of masculine science, *British Journal of the Sociology of Education*, **6**, 133–154

Kelly, A., Whyte, J., and Smail, B. (1984) *Girls into Science and Technology [GIST]: Final Report*. Department of Sociology, University of Manchester

Kelly, A. V. (ed.) (1975) *Case Studies in Mixed Ability Teaching*. London: Harper & Row

Kemmis, S., Atkin, R., and Wright, E. (1977) *How Do Students Learn?* Working Papers on CAL, Occasional Paper No. 5. Norwich: University of East Anglia

Kemp, J. E. and Dayton, D. K. (1985) *Planning and Producing Instructional Media*. New York: Harper & Row

Kempa, R. (1986) *Assessment in Science*. Cambridge: Cambridge University Press

Kerr, J. F. (1963) *Practical Work in School Science*. Leicester: University of Leicester Press

King, D. (1986) 'A note on framework programs', in M. Hope (ed.) *The Magic of the Micro – A Resource for Children with Learning Difficulties.* Newcastle upon Tyne: Council for Education Technology

Klare, G. R. (1974) Assessing readability, *Reading Research Quarterly,* **10**, 62–101

Krathwohl, D. R., Bloom, B. S., and Masia, B. B. (1964) *Taxonomy of Educational Objectives. Handbook II: Affective Domain.* New York: David McKay and Co.

Kuhn, T. S. (1970) *The Structure of Scientific Revolutions.* Chicago, IL: University of Chicago Press

Kyle, W. C. (1980) The distinction between inquiry and scientific inquiry and why high school students should be cognizant of the distinction, *Journal of Research in Science Teaching,* **17**, 123–130

Law, B. (1984) *Uses and Abuses of Profiling.* London: Harper & Row

Lawton, D. (1983) *Curriculum Studies and Educational Planning.* London: Hodder & Stoughton

Levie, W. H. and Lentz, R. (1982) Effects of text illustration: a review of research, *Educational Communication & Technology,* **30**, 195–232

Lindley, J. (1866) *School Botany, Descriptive Botany and Vegetable Physiology; or the Rudiments of Botanical Science.* London: Bradbury, Evans & Co.

Llewellyn-Jones, J. E. (1986) *Body Plans: Animals from the Inside.* Cambridge: Cambridge University Press

Luehrmann, A. (1981) Computer literacy – what should it be? *The Maths Teacher,* **74**, 682–686

Lunzer, E. and Gardner, K. (1979) *Learning from the Written Word.* Edinburgh: Oliver & Boyd

Lusty, M. (1983) Staff appraisal in the education service, *School Organization,* **3**, 371–378

McCall, J., Bryce, T. G. K., and Robertson, I. (1983) Assessing Foundation Science Practical Skills, *Programmed Learning and Educational Technology,* **20**, 11–17

MacDonald, B. (1977) The educational evaluation of NDPCAL, *British Journal of Educational Technology,* **8**, 176–189

MacDonald Ross, M. (1973) 'Behavioural objectives: a critical review', in M. Golby, J. Greenwald, and R. West (eds) (1975) *Curriculum Design.* London: Croom Helm, 355–386

MacFarlane Smith, I. (1964) *Spatial Ability.* London: University of London Press

McNaughton, J. (1986) *Fit for Life.* Basingstoke, Hants: Macmillan

Mager, R. F. (1967) *Preparing Instructional Objectives*. Belmont, CA: Fearon

Malamuth, N. M. and Fitz-Gibbon, C. T. (1977) Tutoring and social psychology: a theoretical analysis, *CSE Report on Tutoring No. 6*. Los Angeles, CA: UCLA

Malone, J. and Dekkers, J. (1984) The concept map as an aid to instruction in science and mathematics, *School Science & Mathematics*, **84**(3), 220–231

Mandelson, R. and Shultz, T. R. (1976) Covariation and temporal contiguity as principles of causal inference in young children, *J. Exptl. Child Psychol.*, **22**, 402–412

Martin, N. (1976) Language across the curriculum: a paradox and its potential for change, *Educational Review*, **2**(3), 206–219

Maslow, A. (1970) *Motivation and Personality*. New York: Harper & Row

Mathews, J. C. (1972) *Teacher's Guide to Assessment in Modern Chemistry*. London: Hutchinson Educational Books

Mayer, V. J. and Richmond, J. M. (1982) An overview of assessment instruments in science, *Science Education*, **66**, 49–66

Medawar, P. B. (1969) *Induction and Intuition in Scientific Thought*. London: Methuen

Merrigan, J. and Herbert, P. (1979) *Basic Skills in Science*. St Albans, Herts: Hart-Davis Educational

Merson, M. W. and Campbell, R. J. (1974) Community education: instruction for inequality, *Education for Teaching*, Spring, 43–49

Midwinter, E. (1972) *Projections: An Educational Priority Area at Work*. London: Ward Lock

Minett, P. M. (1986) *Child Care and Development*. London: Murray

Ministry of Education (1960) *Science in Secondary Schools*, Pamphlet No. 38. London: HMSO

Mitchell, G. C. and Snape, G. W. (1982) *Access to Science*. London: Harrap

Musgrove, F. (1979) *School and the Social Order*. Chichester, Sussex: John Wiley & Sons

NEA [Northern Examining Association] (1986) *Modular Science: Proposals for Examination – 1988*. Manchester: JMB

NFER [National Foundation for Educational Research] (1986) *The Educational Guidance & Assessment Catalogue*. Windsor: NFER/Nelson

Nisbet, S. (1968) *Purpose in the Curriculum*. London: London University Press

Noll, V. H., Scannell, D. P., and Craig, R. C. (1979) *Introduction to Educational Measurement*. Boston, MA: Houghton Mifflin

Novak, J. D., Gowin, D. B., and Johansen, G. T. (1983) The use of concept mapping and knowledge vee mapping with junior high school students, *Science Education*, **67**, 625–645

Nuffield Biology (1966) *Teachers' Guide I: Introducing Living Things*. London: Longman/Penguin Books

Oldroyd, D., Smith, K., and Lee, J. (1984) *School Based Staff Development Activities*. York: Longman/Schools Council

Ormerod, M. B. (1975) Subject preference and choice in co-educational and single sex secondary schools, *British Journal of Educational Psychology*, **45**, 257–267

Osborne, R. and Freyberg, P. (1985) *Learning in Science: The Implications of Children's Science*. London: Heinemann Educational Books

Osborne, R. and Wittrock, M. (1985) The generative learning model and its implications for science education, *Studies in Science Education*, **12**, 59–87

Pankratz, R. (1967) 'Verbal interaction patterns in the classrooms of selected physics teachers', in E. J. Amidon and J. B. Hough (eds) *Interaction Analysis: Theory, Research and Application*. Reading, MA: Addison-Wesley

Paskell, T. (1985) When the wind blows, *Natural World*, Autumn, 14–15

Pearson, H. and Wilkinson, A. (1986) The use of the word processor in assisting children's writing development, *Educational Review*, **38**, 169–187

Peck, M. J. and Williams, J. P. (1978) Science for the least able pupils leading to a C.S.E. qualification, *School Science Review*, **60**, 353–357

Perera, K. (1980) The assessment of linguistic difficulty in reading material, *Educational Review*, **32**, 151–161

Perera, K. (1984) *Children's Writing and Reading: Analysing Classroom Language*. Oxford: Blackwell

Piaget, J. (1950) *The Psychology of Intelligence*. London: Routledge & Kegan Paul

Popham, W. J. (1970) 'Probing the validity of arguments against behavioral goals', in R. J. Kible, L. L. Barker, and D. T. Miles, *Behavioral Objectives and Instruction*. Newton, MA: Allyn & Bacon

Popham, W. J. and Baker, E. L. (1970a) *Systematic Instruction*. Englewood Cliffs, NJ: Prentice-Hall

Popham, W. J. and Baker, E. L. (1970b) *Establishing Instructional Goals*. Englewood Cliffs, NJ: Prentice-Hall

Price, J. and Talbot, B. (1984) Girls and physical science at Ellis Guilford School, *School Science Review*, **66**, 107–111

Reid, D. J. (1980) Spatial involvement and teacher–pupil interactional patterns in school biology laboratories, *Educational Studies*, 6(1), 31–41

Reid, D. J. (1984a) A three-in-one readability program for science worksheets, *School Science Review*, 65, 560–569

Reid, D. J. (1984b) Readability and science worksheets in secondary schools, *Research in Science and Technological Education*, 2(2), 153–165

Reid, D. J. (1984c) The picture superiority effect and biological education, *Journal of Biological Education*, 18(1), 29–36

Reid, D. J. and Beveridge, M. (1986) Effects of text illustration on children's learning of a school science topic, *British Journal of Educational Psychology*, 56, 294–303

Reid, D. J., Beveridge, M., and Wakefield, P. (1986) The effect of ability, colour and form on children's perception of biological pictures, *Educational Psychology*, 6(1), 9–18

Reid, D. J., Briggs, N., and Beveridge, M. (1983) The effect of picture upon the readability of a school science topic, *British Journal of Educational Psychology*, 53, 327–335

Reid, D. J. and Miller, G. J. A. (1980) Pupils' perception of biological pictures and its implication for readability studies of biology textbooks, *Journal of Biological Education*, 14(1), 59–69

Reid, D. J. and Shields, C. (1985) Microcomputers and biology software – identifying a problem, *Research in Education*, 34, 1–8

Reid, D. J. and Tracey, D. C. (1985) The evaluation of a school science syllabus through objectives and attitudes, *European Journal of Science Education*, 7, 375–386

Resources for Learning Development Unit (1985) *Science for the Individual*. Bristol: RLDU

Revised Nuffield Biology (1974) *Teachers' Guide I*. London: Longman

Revised Nuffield Chemistry (1975) *Teachers' Guide I*. London: Longman

Revised Nuffield Physics (1977) *Teachers' Guide*. London: Longman

Richardson, E. (1967) *The Environment of Learning*. London: Nelson

Roberts, D. A. (1982) Developing the concept of 'curriculum emphases' in science education, *Science Education*, 66, 243–260

Rockefeller Report (1958) *The Pursuit of Excellence, Education and the Future of America*. Garden City, NY: Doubleday

Rose, C. D. (1981) Social integration of school age ESN(S) children in a regular school, *British Journal of Special Education: Forward Trends*, 8(4), 17–22

Rosenthal, R. and Jacobsen, L. (1968) *Pygmalion in the Classroom: Teacher Expectation and Pupils' Intellectual Development*. New York: Holt, Rinehart & Winston

Rostron, A. and Sewell, D. (n.d.) *Microtechnology in Special Education*. London: Croom-Helm

Rotter, J. B. (1966) Generalized expectancies for internal versus external control of reinforcement, *Psychological Monographs*, **80**, No. 609

Rowe, M. B. (1974a) Wait-time and rewards as instructional variables their influence on language, logic, and fate control: part I – wait-time, *Journal of Research in Science Teaching*, **11**(2), 81–94

Rowe, M. B. (1974b) Relation of wait-time and rewards to the development of language, logic, and fate control: part II – rewards, *Journal of Research in Science Teaching*, **11**(4), 291–308

Rowntree, D. (1977) *Assessing Students: How Shall We Know Them?* London, Harper & Row

Royal Society (1982) *Science Education 11–18*. London: Royal Society

Royal Society (1985) *The Public Understanding of Science*. London: Royal Society

Rubowits, P. C. and Maehr, M. L. (1971) Pygmalion analysed: toward an explanation of the Rosenthal–Jacobson findings, *Journal of Personality & Social Psychology*, **19**, 197–203

Rushby, N. J. (1978) *An Introduction to Educational Computing*. London: Croom Helm

Samuels, S. J. (1970) Effects of pictures on learning to read, comprehension and attitudes, *Review of Educational Research*, **40**, 397–407

Sands, M. K. and Bishop, P. E. (1984) *Practical Biology: A Guide to Teacher Asessment*. London: Bell & Hyman

Satterly, D. (1981) *Assessment in Schools*. London: Basil Blackwell

Scheflen, A. & Scheflen, A. (1975) *Body Language and Social Order*. Englewood Cliffs, NJ: Prentice-Hall

Schibeci, R. A. (1984) Attitudes to science: an update, *Studies in Science Education*, **11**, 26–59

Schools Council (1973a) *Assessment of Attainment in Sixth Form Science*, Schools Council Exam. Bull. No. 27. London: Evans/Methuen

Schools Council (1973b) *Science 5–13*. London: Macdonald

Schools Council (1980) *Learning Through Science*. London: Macdonald

Schools Council (1983) *Science for Children with Learning Difficulties*. London: Macdonald

Schwab, J. J. (1964) Structure of the disciplines: meanings and significances, in A. W. Ford and L. Pugno (eds) *The Structure of Knowledge and the Curriculum*. Chicago, IL: Rand McNally, 6–30

SCISP (1974) *Schools Council Integrated Science Project*. London and Harmondsworth, Middx: Longman/Penguin

Scriven, M. (1967) 'The methodology of evaluation', in R. E. Stake (ed.) *Perspectives in Curriculum Evaluation*, AERA Monograph No. 1. Chicago. IL: Rand McNally, 39–89

SEC [Secondary Examinations Council] (1986) *Science G.C.S.E. A Guide for Teachers*. Milton Keynes: Open University Press

Seligman, M. E. P. (1975) *Helplessness*. San Francisco: W. H. Freeman

SEMERC [Special Education Microelectronic Resources Centre] (1986) *Supporting the Slow Learner – the Power of the Micro*. Manchester: Council for Educational Technology

Shapin, S. and Barnes, B. (1976) Head and hand: rhetorical resources in British pedagogical writing, *Oxford Review of Education*, **2**, 231–254

Shayer, M. and Adey, P. (1981) *Towards a Science of Science Teaching*. London: Heinemann

Skilbeck, M. (1982) *A Core Curriculum for the Common School*. London: University of London Institute of Education

Skilbeck, M. (ed.) (1984) *Evaluating the Curriculum in the Eighties*. London: Hodder & Stoughton

SMA [Science Masters Association] (1957) *Science and Education: A Policy Statement by the Committee of the SMA*. London: John Murray

Smail, B. (1985) *Girl-friendly Science: Avoiding Sex Bias in the Classroom*. York: Longman/Schools Council

Smail, B. and Kelly, A. (1984a) Sex differences in science and technology among eleven year old schoolchildren: I cognitive, *Research in Science & Technological Education*, **2**, 61–76

Smail, B. and Kelly, A. (1984b) Sex differences in science and technology among eleven year old schoolchildren: II affective, *Research in Science & Technological Education*, **2**, 87–106

Smith, G. C. and Curtis, S. (1986) *It's Your Life*. York: Longman Resources Unit

Smith, V. (1983) Teaching science to the slow learner, *School Science Review*, **65**, 138–140

Solomon, J. (1980) *Teaching Children in the Laboratory*. London: Croom Helm

Solomon, J. (1983) Messy, contradictory and obstinately persistent: a study of children's out of school ideas about energy, *School Science Review*, **65**, 225–229

Sommer, R. (1969) *Personal Space: The Behavioral Basis of Design*. Englewood Cliffs, NJ: Prentice-Hall

Sparkes, R. A. (1983) *The BBC Microcomputer in Science Teaching*. London: Hutchinson

Special Needs Computer Centre (1986a) Using Prompt 2 in biology, *Newsletter No. 3.* Manchester: Council for Educational Technology

Special Needs Computer Centre (1986b) 'Prompt 3: A supportive introductory word-processor'. Manchester: Council for Educational Technology

SSCR [Secondary Science Curriculum Review] (1983) *An Outline of the Purpose, Organization and Operation of the Review.* London: Schools Council

SSCR (1984) *Towards the Specification of Minimum Entitlement: Brenda and Friends.* London: Schools Council

Stenhouse, L. (1975) *An Introduction to Curriculum Research and Development.* London: Heinemann

Stillman, A. B. (1982) The rationale for abilities testing and specifically the development of a scientific ability test, *School Science Review,* **63**, 423–433

Suppes, P. (1966) The uses of computers in education, *Scientific American,* **215**(2), 206–220

Sutton, C. (ed.) (1981) *Communication in the Classroom.* London: Hodder & Stoughton

Sutton, R. (ed.) (1986) *Assessment in Secondary Schools: The Manchester Experience.* York: Longman/School Curriculum Development Committee

Swedish Ministry of Agriculture (1982) *Acidification Today and Tomorrow.* Stockholm: Swedish Government

Taba, H. (1962) *Curriculum Development: Theory and Practice.* New York: Harcourt Brace & World

Taber, F. M. (1983) *Microcomputers in Special Education.* Heston, VA: Council for Exceptional Children

Tawney, D. (ed.) (1976) *Curriculum Evaluation Today: Trends and Implications.* London: Macmillan/Schools Council

Taylor, J. (ed.) (1979) *Science at Work.* London: Addison-Wesley

Thelen, H. A. (1969) Tutoring by students, *School Review,* **77**, 229–244

Torrance, E. P. (1974) *Torrance Tests of Creative Thinking.* Lexington, MA: Ginn & Co.

Travers, N. (1984) *The Teaching of Science to Pupils of Low Educational Attainment.* Northumberland County Council

Tyler, R. W. (1949) *Basic Principles of Curriculum and Instruction.* Chicago, IL: University of Chicago Press

Vachon, M. K. and Haney, R. E. (1983) Analysis of concepts in an eighth grade science textbook, *School Science and Mathematics,* **83**(3), 236–245

Van Praagh, G. (1973) *H. E. Armstrong and Science Education*. London: Murray

Vygotsky, L. S. (1962) *Thought and Language*. Cambridge, MA: MIT Press

Walford, G. (1983) Science textbook images and the reproduction of sexual divisions in society, *Research in Science and Technological Education*, **1**, 65–72

Walker, R. and Adelman, C. (1975) *A Guide to Classroom Observation*. London: Methuen

Ward, C. (1980) *Designing a Scheme of Assessment*. Cheltenham, Glos.: Stanley Thornes

Watkins, O. (1981) 'Active reading and writing', in C. Sutton (ed.) *Communication in the Classroom*. London: Hodder & Stoughton

Watts, D. M. and Zylberstajn, A. (1981) A survey of some children's ideas about force, *Physics Education*, **16**, 360–365

Weiner, B. (1979) A theory of motivation for some classroom experiences, *Journal of Educational Psychology*, **71**, 3–25

Wheeler, D. K. (1967) *Curriculum Process*. London: University of London Press

Whitfield, R. C. (1979) Educational research and science teaching, *School Science Review*, **60**, 411–430

White, J. P. (1971) The concept of curriculum evaluation, *Journal of Curriculum Studies*, **3**, 101–112

White, R. T. (1979) Achievement, mastery, proficiency, competence, *Studies in Science Education*, **6**, 1–22

Wicks, S. (1986) 'Colour combinations presented on a microcomputer'. M.Ed. dissertation, University of Manchester

Wilkinson, A. L. (1983) *Classroom Computers and Cognitive Science*. London: Academic Press

Wilkinson, D. and Bowers, J. (1976) *LAMP Project. A Teachers' Handbook I. A Guide to the Project*. Hatfield, Herts: Association for Science Education

Witkin, H. A. (1975) 'Some implications of research on cognitive style for problems in education', in J. M. Whitehead (ed.) *Personality and Learning I*. London: Hodder & Stoughton

Woolnough, B. E. and Allsop, T. (1985) *Practical Work in Science*. Cambridge: Cambridge University Press

Young, B. L. (1979) *Teaching Primary Science*, Harlow, Essex: Longman

Young, M. F. D. (1976) 'The schooling of science', in G. Whitty and M. F. D. Young (eds) *Explorations in the Politics of School Knowledge*. Driffield, N. Humberside: Nafferton Books

Name Index

Subject Index